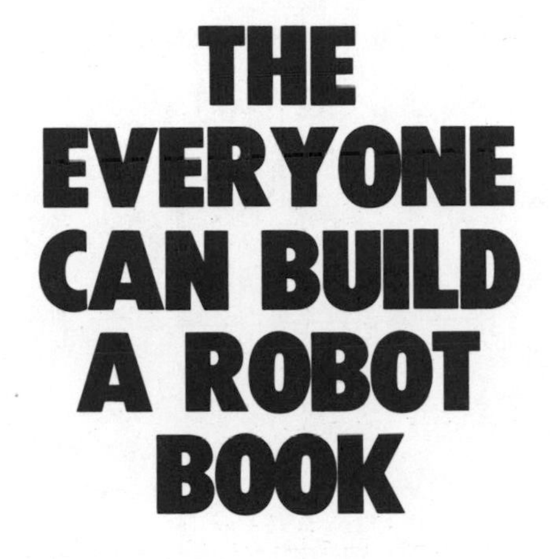

THE EVERYONE CAN BUILD A ROBOT BOOK

Other Books by Kendra Bonnett and the Editors of
DIGIT Magazine:

Ace Your Grades with Your Computer (Jan. pub)

THE EVERYONE CAN BUILD A ROBOT BOOK

Kendra R. Bonnett and Gene Oldfield

Editor
Howard Rheingold

Coordination Editor
Lassie Benton

Illustration by Bob Johnson

The Computer Book Division
Simon & Schuster, Inc.
New York

Published by
the Computer Book Division/Simon & Schuster, Inc.
Simon & Schuster Building
Rockefeller Center
1230 Avenue of the Americas
New York, New York 10020
SIMON AND SCHUSTER and colophon are registered trademarks
of Simon & Schuster, Inc.

Designed by Design Vectors, San Francisco, CA
Manufactured in the United States of America

10 9 8 7 6 5 4 3 2 1

Library of Congress Cataloging in Publication Data

Bonnett, Kendra.
The everyone can build a robot book.

1. Robots—Design and construction. I. Oldfield,
Gene. II. Digit magazine. III. Title.
TJ211.B66 1984 629.8'92 84-19876
ISBN: 0-671-53059-3

THE EVERYONE CAN BUILD A ROBOT BOOK

CONTENTS

DEDICATION

Mr. Bayne Kelley—
I am keeping a promise that I made to myself several years ago, to dedicate my first book to the teacher who taught me to appreciate the English language. I only wish I had known at the time what a wonderful gift you were giving me. In case I forgot to say it then, thank you.

ACKNOWLEDGMENTS

No matter how self-reliant and independent authors may be, they still need the assistance and advice of others. We wish to thank Howard Rheingold for his editorial advice. Dovell M. Bonnett worked long hours for very little pay, helping his sister begin to understand electronics. He also designed the Experimenter's Interface Board and drew the schematics. Finally, Charles Lichtenstein, an electronics instructor at San Jose State University, provided helpful suggestions regarding the Moth and introduced us to optocouplers.

INTRODUCTION

If you want to build a robot, read on. By the time you finish this book, you will be able to build your own working robots and you will know something about the basic principles of electronics and robotics (the science of robots). We're not talking about building just another mechanical toy, but a working robot, complete with photosensory feedback. Best of all, you do not need an advanced degree in electrical and mechanical engineering or a big budget to do it. If you can read, follow directions, and use a few simple tools, you *can* build a robot.

The Everyone Can Build a Robot Book is more than a cookbook for building a robot. The first half gives you some background on robots and explains the basic principles of electronics. The second half contains the projects. We recommend that you read the book in the order it is laid out. But if you just cannot wait to start and have a working knowledge of electronics, then turn to page 45. Or, if you feel that you need a little background in electronics first, turn to page 31.

Only a few years ago, any book about robots would have been filed in the science-fiction section of a library or bookstore. But while Hollywood's Robby the Robot was a fake, real robots are working in factories, in outer space, under the sea, and are even running around in people's homes at this very minute. And robotics is still in its infancy. Today's robots are like Model T's. During the next ten years or so, Topos, RB5Xs, homemade robots, and industrial robot arms will evolve into the Ferraris of the robot world. The fantastic machinery of yesterday's science fiction will become as commonplace as telephones, televisions, and toasters.

Launch your study of robotics by taking a trip into the past, where you will look at robots in myth and history. Did you know that the modern robot was conceived in the imaginations of Egyptian and Greek storytellers? Even the great inventor Leonardo da Vinci dabbled in robot design.

Moving forward in time, you will study the development of the modern robot—both the Hollywood version and the industrial model. You will discover the role robots are beginning to play in our homes. You may well buy a personal robot in the next five to ten years! Today's home robots may be little more than high-tech pets, but tomorrow they may become maids, cooks, and babysitters rolled into one, and certainly a sentry, entertainment, and communication center.

Once you understand the robot's form and function and how the whole idea of robots developed, you can start getting into the technical material. You will learn some of the elements of electronic circuitry involved in robotics. Then, you can begin building your own robots from inexpensive parts that are easy to find around your house, a hardware store, and your local Radio Shack. Most important, you'll learn about the essential concepts of homebrew robot design. By learning how to *think* about these devices—what they might become as well as what they are, what they can do as well as can't—you will be launching yourself on a new and rewarding hobby, and maybe even an exciting career in robot design.

You will construct a robot called the Moth. It is a small robot with a light sensor that enables you to control its movements with a flashlight. Once you have built the basic Moth, you will learn how to customize it and use it as the basis for automating toys and models. By the time you finish reading our suggestions, you'll probably have several ideas of your own! If these projects whet your appetite for more, we're also going to get you started on more advanced robot projects that will help you connect your robot to a home computer!

With that, let's get started. Just remember that for all its roots in imagination and myth, robotics is a real science that is playing an increasingly important role in our modern world. As a robot builder, you may play a part in its development.

SCROUNGERS' SHOPPING LIST

Even before you get to the projects, you should start gathering the tools and spare parts that every robot builder needs. Many of the items are available around the house. Others can be scrounged at garage sales and flea markets or salvaged from broken toys. When it comes time to build the robot Moth, you will need to buy a few parts at your local Radio Shack store.

1. Don't throw anything away! A good homebrew robot builder is also a packrat. We guarantee that the minute you throw something away you will wish you had it. To keep your work area from looking like a junk shop, devise some sort of filing system. Store spare parts in shoe boxes and label the boxes.

2. Toy gearhead motors—You can scrounge 3-6 VDC (volts DC) motors from battery-operated toys. *The best ones are in radio-controlled toys.* The motor, gear and axle can be removed as a unit and mounted directly on your robot! Hobby and electronics stores also carry little motors. As you become more ambitious and decide to tackle larger problems, you may need other kinds of motors. (Car windshield wiper motors are particularly good for larger robots.) If you should need special motors or want to know more about motors, order a free copy of the *Motion Control Handbook* from B&B Motor and Control Corporation, Corporate Marketing and Communications, Rte. 4 and Covey Road, Burlington, CT 06013.

3. Wheels and tires—Once again, toys are your best (and cheapest) source. If you want tires with a little more bounce or larger diameter, try hobby, airplane tires. If you can't scrounge these, try a hobby store. They will cost about $2.50/pair.

4. Nuts & Bolts—It's a good idea to keep an assortment of nuts, bolts, and wood screws on hand. You never know what your going to need when you start creating.

5. Electronic components—An assortment of resistors, disk capacitors, and diodes are a must. You can get great deals when you shop at computer and electronics flea markets and swap meets. Also, Radio Shack can meet most of your needs, especially when you are getting started. If you are not sure what to buy, but want to start collecting parts, look ahead to the basic electronics section (page 31) and at the shopping list for the Moth (page 46). In your spare time, you might want to sort the resistors by ohms. This will save you time later.

6. Wood and plastic scrap—Odds and ends (also called remainders) of wood and plastic come in handy for building robot platforms. Wood is usually available around the house, or you can pick up scraps at a lumberyard or hardware store. You might also try the woodworking shop at school. If you prefer plexiglas bases, check out the scrap bin at your local plastic supply store. The pegboard in the Lite Brite toys make excellent robot platforms— the holes are already drilled.

7. Tools—You can begin assembling the tools you will need for building a robot. Some of the more common tools are probably available around the house, such as hammer, files, screwdrivers, hacksaw, and electric drill. Then turn to page 38 for a detailed list. A good robot builder needs the right tools.

8. Future Equipment—As you progress into larger, more sophisticated robot projects, you will want to keep an eye out for large, rideable, motorized toys. They make excellent bases for robots.

MYTHS & IMAGES—MECHANICAL BEINGS FROM ANCIENT GREECE TO HOLLYWOOD

Recall, for a moment, that far-off galaxy. Our heroes are trapped deep in the bowels of Lord Vader's Death Star, about to be crushed in the ship's giant trash masher. Their cunning and skill are no match against the powerful walls closing in around them. Their only hope is that a couple of robots, C-3PO and R2D2, can break the code of the ship's computer, disable the Death Star, and stave off disaster down in the garbage pit.

Both R2D2 and C-3PO are highly-developed, intelligent machines, capable not only of responding to human commands but of acting on their own. They even have distinct personalities. R2D2 is cute, sort of a space-age teddy bear, and C-3PO, while somewhat clumsy, is a devoted friend and companion.

Intelligent, sophisticated examples of high technology, almost human, and sometimes dangerous and evil—this is how science-fiction movies, novels, and television have portrayed robots. Most of our expectations for robots—how they should look and act—have been shaped by fantasy and Hollywood. One of the earliest examples was Robby the Robot from a 1956 film entitled *Forbidden Planet.* Robby, incidentally, was also one of the few friendly movie robots.

While the robot is a product of both science fiction and the very specialized science of robotics, ours is not the first generation to envision machines that look and function like humans. The word "robot" (which comes from the Czechoslovakian word "robotta'—meaning degrading or menial work) was first used in 1921 to refer to the mechanical workers in the play *R.U.R. (Rossum's Universal Robots).* But the idea is older, dating back at least to ancient Egypt and Greece.

The earliest stories of inanimate objects being endowed with characteristics of man and beast could be called supernatural or mystical (even mythical) phenomena. For example, an Egyptian statue of the slain Memnon—victim of the Trojan War—was said to speak when struck by the first rays of dawn's light. Its voice was "like that of a harp or lyre with a broken chord."

According to myth, the Greek inventor Daedalus created statues outside the Labyrinth on Crete that were so lifelike they had to be chained to keep them from running away. The Greeks thought that these statues were somehow powered by quicksilver (Mercury).

Stories of mechanical people grew increasingly popular in the Middle Ages and Renaissance. The English friar and philosopher, Roger Bacon, is supposed to have made a talking head. A young friar was asked to watch the head and report to Bacon if anything happened. After everyone else left the room, the head said, "The time is now," indicating that the man should call Roger Bacon. The friar ignored the head, which spoke again, "The time is now." Still the brother ignored it. Then the head said, "The time is past," and fell silent forever.

Thirteenth-century inventor Albertus Magnus spent 20-30 years of his life creating an automated being out of glass, metal, wood, wax, and leather. Thomas Aquinas, later canonized as a Catholic saint, destroyed the mechanical man, declaring it to be the work of the Devil!

WOULD YOU BELIEVE . . .

The ancients were captivated with the idea of creating inanimate, mechanical objects that resembled living creatures. History is full of reports of inventors creating a wooden pigeon, powered by water, that appeared to fly (Greece, 400 B.C.); a wooden horse that bounded along on springs and a bamboo magpie that flew (China, 500 B.C.); a wooden otter that caught fish (China, 790 A.D.); and a rat-catching, wooden cat (China, 890 A.D.).

In fifteenth-century Germany, Johannes Muller created a mechanical eagle that reportedly flew into the city of Nuremberg, landed on a gate atop the city, and greeted Holy Roman Emperor Maximilian I by stretching its wings and bowing. Similarly, Leonardo da Vinci developed an automated lion to honor King Louis XII of France upon the latter's visit to Milan. The lion is supposed to have approached Louis, reared up, and opened a compartment in its breast, revealing a fleur-de-lis, France's royal coat of arms.

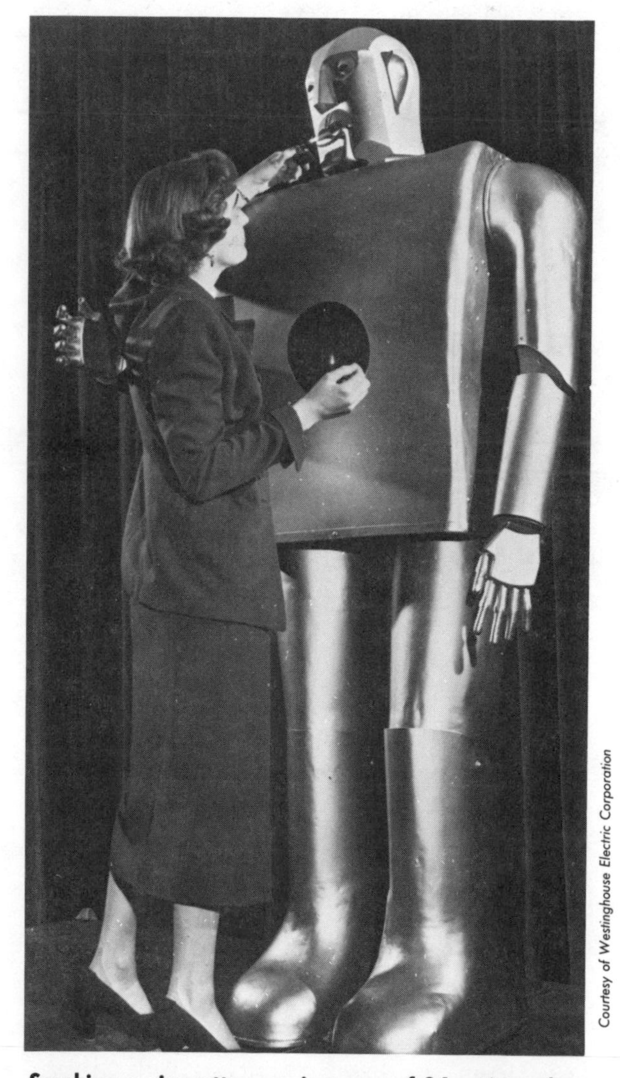

Courtesy of Westinghouse Electric Corporation

Smoking a cigarette was just one of 26 actions that made Elektro the hit of the 1939 World's Fair.

The response was quite different at the 1939 New York World's Fair as crowds swarmed around the Westinghouse Electric exhibit trying to catch a glimpse of Elektro and a mechanical dog named Sparko. Elektro moved its fingers and arms, walked (rolled), and produced great puffs of smoke—twenty-six movements in all. The faithful Sparko barked, begged, and wagged its tail. But even these amazing characters were only remote-controlled machines designed to perform a series of predetermined movements—not robots. The machines were voice activated. They responded to a series of two-word (to start a movement) and one-word (to stop) commands, which were interpreted as vibrations and converted into electrical impulses, activating the appropriate response.

All these stories of mechanical creatures sound more like something from *That's Incredible* or *Ripley's Believe It or Not,* than an historical account of robots. While we do not have to believe that our ancestors were remarkable robot builders, we should recognize in their primitive tinkering a trait still present today. It is our fascination, even obsession, with duplicating ourselves in a form that can live forever. Some call it a desire to "play God." Others say it is an attempt to deny our mortality. Rather than "cheating death," robot builders are probably more interested in creating life.

Stories, such as that of the Frankenstein monster, tell us that meddling with nature and God can get us into trouble. Today, with technology more powerful, more potentially destructive than ever before, this message is reinforced constantly. Scientists and engineers have the power to help create life (biogenetics) as well as take it away (nuclear war). A television commercial a few years back told us that "it's not nice to fool with Mother Nature." While that may be a bit extreme, you, as a robot builder, need to consider the consequences of your work. For example, will your creation help improve the image of robots or will it just end up reinforcing old myths?

Hollywood, more than any other medium, has shown us the dark side of robotics. Thanks to movies like *The Day the Earth Stood Still* (1951), *Silent Running* (1971), *West World* (1973), and *Logan's Run* (1976) and television programs like *Lost in Space* and *Star Trek,* we have seen people do battle with these intelligent, sophisticated examples of high technology. These robots have appeared almost human—in fact super human—and quite often dangerous and evil. The media have encouraged us to have a love-hate relationship with robots (with the emphasis on the hate). We love the idea that they can care for us and do the work that we do not like, but we hate and fear their potential for destruction and harm. In fact, it has only been since the *Star Wars* epic and the introduction of C-3PO and R2D2 that we have begun to soften our attitude.

Science fiction has had such an adverse effect on people's attitudes toward robots that back in 1940 Isaac Asimov drew up the Three Laws of Robotics, which he introduced in a short story entitled "I Robot." Asimov, a scientist, prolific science-fiction writer and futurist, was disturbed by the number of stories written about robots destroying their creators. He established these laws for robot design in order to diminish people's fears that robots would harm them. The laws are as follows:

1. *A robot may not injure a human being, or, through inaction allow a human being to come to harm.*
2. *A robot must obey the orders given it by human beings except where such orders would conflict with the First Law.*
3. *A robot must protect its own existence as long as such protection does not conflict with the First or Second Law.*

Illustration by Bob Johnson

While sound in principle, many authors have been slow to give up the old image of the dangerous, uncontrollable robot. One reason is the apparent loophole in Asimov's laws. What if the robot's microprocessor is damaged and the robot begins malfunctioning? We've all seen computer programs do things they are not programmed to do, why should robots be different? The answer is, they are not! In fact, a story made the headlines recently about a Japanese industrial worker who was killed by a robot. In this case, the robot did not go berserk; the human simply got in the way of the arm. The robot swung around to pick up a piece of equipment and struck the man with a deadly blow, proof that things can go wrong.

As a robot builder, you must understand these human prejudices and concerns regarding robots and strive to overcome them. One way is to examine carefully what movies, television, and science fiction have done to our perception of robots. If you can determine the physical characteristics, for

Illustration by Bob Johnson

example, that appeal to people and design your robots accordingly, your creations will enjoy greater acceptance. Similarly, if your robot is going to be working on the ocean floor or in outer space—someplace where it will not be in contact with a lot of people—then it may not be necessary to place so much emphasis on appearance.

To get you started, let's examine the robot as portrayed in fiction. There are two basic shapes. One is the android design, which closely resembles the human figure. Take a look at C-3PO. The second looks more like a machine, often resembling a canister vacuum cleaner. R2D2 fits this latter description. Unless they roll around on wheels, fictional robots traditionally walk upright, like humans. Actually, they lurch from side to side, rather like the Frankenstein monster! (You might want to avoid that connection.) Robots have eyes with x-ray vision, they talk through little speakers cleverly placed where a mouth should be.

All this fanciful design is perhaps just one more example of man trying to create life in his or her own image, but the fact remains that it is an image that has served us for a long time. Besides, it's an image that people recognize. As a robot builder, you should keep this in mind, at the same time keeping yourself open to fresh ideas. There is, after all, no functional reason that robots need to resemble the human form. Look at the industrial model—the robot arm. This is the most productive robot in existence. It moves along six axes, duplicating basic human movements. That's where the similarity ends, for it certainly does not look like anyone you've seen walking down the street! Suffice to say that if you can humanize your robot creations without sacrificing efficiency and economy, that's success.

GOLEMS

Starting in the Middle Ages, Jewish folklore began to tell of strange, unformed monsters—called golems. Golems were created to serve; their masters brought them to life through strange, religious rituals. While their general appearance was humanoid, they were misshapen and incompletely formed. Golems were as tall as a full-grown adult, but they supposedly resembled the fetus of an unborn child, sort of embryonic monsters. In legend, Ben Sira wrote the Hebrew word for truth, "emeth," on the forehead of a golem, thereby giving it life. Gustav Meyrink's novel, *The Golem,* is the story of a frenzied creature that beseiged the city of Prague. As with the story of Frankenstein, the rabbis who created golems were not evil. Rather, they were men dabbling in the powerful science of creation. In the end, they and their creations were destroyed.

HUMANOID describes a creature with human characteristics. A Martian that had two arms and legs, pointed ears, red skin, and walked upright like an earthling would be a humanoid.

THIS BIRD HAS GOOD TIMING

So far you have read some rather strange and farfetched stories about early robots. We have to wonder about the truth in these ancient tales. As we know, modern robots can scarcely perform some of the more intricate tasks attributed to those early machines. On the other hand, mechanical creations, such as the Strasbourg cathedral rooster, come closer to reality. People came from far and wide to watch the mechanical bird flap its wings and crow as the cathedral clock struck the hour.

MAKE MINE A SHORT ONE WITH NO ARMS

A group of people who had grown up watching the terrifying science-fiction movies of the fifties and sixties were asked what they thought about having robots in their homes. Almost all agreed that they liked the idea of mechanical servants, but they had some criteria for their design: 1) Robots must not be too tall; in the movies huge robots have dominated people. 2) Robots should not have hands; they might get out of control, pick up a knife, and start killing.

2

MACHINES IN OUR IMAGE

Having seen movies like *Star Wars* and watched the amusing routines of so-called remote-controlled, entertainment robots (called showbots), many people assume that the technology is at the point where robots can converse with humans and attend to their needs. When they see a personal robot, such as RB5X or HERO 1, they are disappointed. Their first response is seldom, "Wow, that is incredible . . . the sonar prevents the robot from banging into the wall." Rather, they say, "But what else can it do?"

Courtesy of World of Robots

Like most showbots, Denby is friendly-looking and unintimidating.

If you use fictional robots as the model, real robots are disappointing. Personal robots, for example, are limited to entertaining their owner with songs and dances, turning on appliances, serving as a mobile burglar alarm and smoke detector, and answering the phone. A few can vacuum the floor, but don't expect them to clean behind the couch.

Industrial robot arms are more productive. They represent a direct replacement for human labor, serving as auto painters, parts inspectors, welders, assemblers, and materials handlers. Although they look nothing like humans, robot arms simulate several basic human characteristics. Their gears and motors work like muscles. Most robot arms have six axes (directions) of movement—arm sweep, shoulder swivel, elbow extension, and three wrist motions. Their electronic circuits function like nerves; light-sensitive cells provide a sense of touch; and television scanners, sonar, and new image-recognition systems serve as eyes. But what makes robots special, distinguishing them from all other machines, is their computer "brain" and memory capacity. Robots can be programmed and reprogrammed to do a variety of tasks—making them more versatile than conventional machinery. Take a toaster, for example, it's great for making toast. But don't expect it to sweep floors—you need a vacuum. Someday, however, domestic robots will be programmed to make toast, vacuum floors, and much more.

In the meantime, for all their versatility,

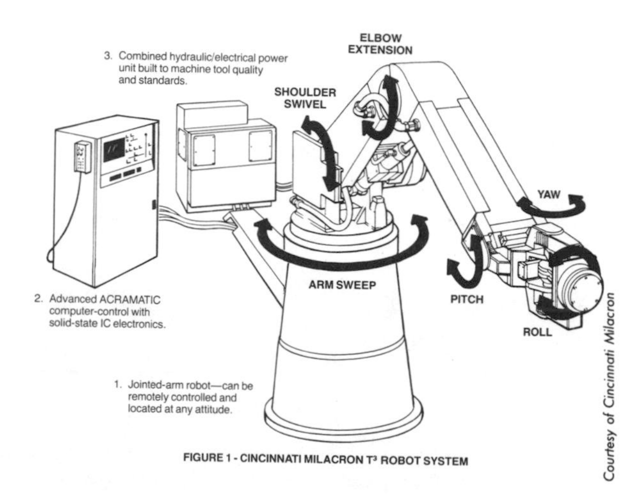

Courtesy of Cincinnati Milacron

Most robot arms have six axes of movement—arm sweep, shoulder swivel, elbow extension, and three wrist motions.

robots still fall far short of the real thing—people. Researchers strive to improve robot motion, artificial intelligence, vision, and speech—and little by little they are making progress. The fact remains, however, that performing a simple task, such as riding a bicycle, would require a major breakthrough in hardware and software development. (Actually, riding a tricycle would be considerably easier.)

The computer—the element that distinguishes robots from other machines—also limits what robots can do. As you know, computers can remember anything programmed into memory. Yet they cannot think or make decisions about anything not in memory. We expect that with advances in artificial intelligence some day computers will begin to think.

In the meantime and until robots demonstrate improved artificial intelligence, mobility, and senses of touch and vision, we cannot turn them loose in our homes and expect them to make dinner, walk the dog, feed the goldfish, vacuum the living room, and make the beds. These tasks, no matter how simple they seem, require complex thought patterns, excellent vision, superior mobility, and keen reflexes.

Suppose that while a robot is vacuuming it comes upon a tennis racket, one skate, and a book all lying on the floor where they don't belong. Now, the robot could be programmed to stop when it found an unfamiliar object and say, "Put your stuff away!" But what if no one is home? It cannot put the tennis racket in the hall closet, find the other skate and put them both in the toy bin, reshelve the book, and finish vacuuming.

Similarly, if all the ingredients were laid out, a robot with an advanced image-discrimination system could probably be programmed to make brownies. It could also turn on the oven, put the pan in the oven,

WHAT IS A ROBOT?

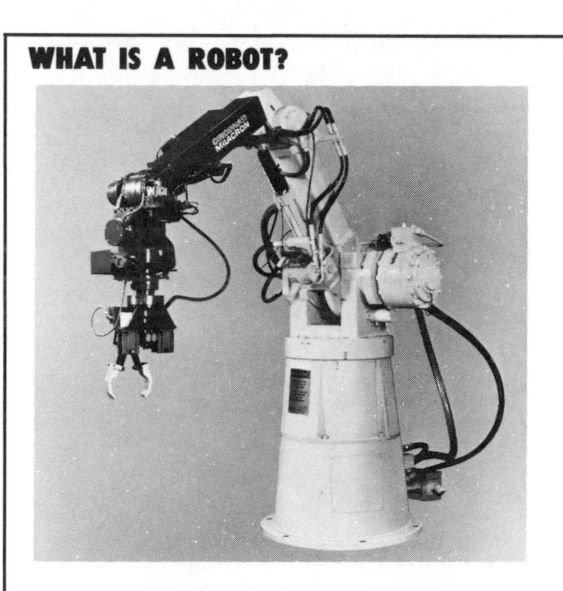

Courtesy of Cincinnati Milacron

"A robot is a reprogrammable, multifunctional manipulator designed to move material, parts, tools, or specialized devices through variable programmed motions for the performance of a variety of tasks."

—*Robot Institute of America*

"I don't quite know how to define one, but I know one when I see it."

—*Joseph Engelberger, founder of Unimation*

and take it out in 20 minutes. But what if you told the robot to make brownies, and you had not checked to see that all the ingredients were there? The robot goes to the refrigerator, reaches for the milk, and finds, instead, an empty carton. No robot could stop what it is doing, drive (or roll) down to the store, buy milk, come home, and finish cooking.

Both of these simple, everyday tasks require faculties that robots do not have—yet. Scientists, engineers, computer specialists, and homebrew designers in the area of robotics are working to solve the problems.

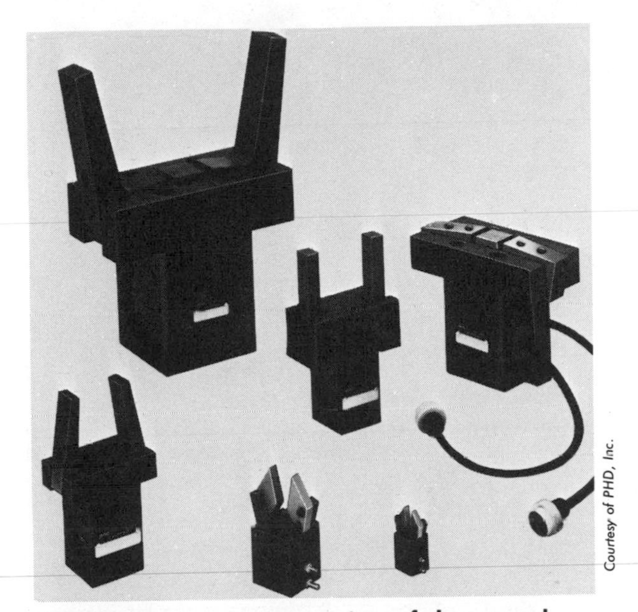

Courtesy of PHD, Inc.

Robot grippers come in a variety of shapes and sizes. Some have vice-like grips. Others rely on magnets or suction cups.

Expect improved vision, a refined sense of touch, and greater mobility to come first. While these are not simple problems, they are more questions of electrical and mechanical technology. Scientists are doing some interesting work with a new robot skin that will heighten the sense of touch.

Intelligence is both one of the more difficult and interesting subjects that a robot builder can study. How do human beings think? Scientists and philosophers alike are trying to understand human intelligence. This study and the next step (making machines replicate human behavior) are called artificial intelligence, or AI for short. Artificial intelligence is a vast subject, one that is largely unexplored and boasting few major breakthroughs.

One development that you may be familiar with is the computer language Logo. Researchers at the Massachusetts Institute of Technology have been using Logo with preschool children to study natural or native intelligence (that behavior we are born with). Because Logo is both powerful and easy to learn, even young children can take charge of the computer and direct their own learning. Scientists observe how these children manipulate the computer and use Logo in the learning process.

In spite of the research, scientists have been able to give robots only the most limited power to think. Most of that 'thinking'' is of the trial and error variety. Certain rules are programmed into the computer. Some of these rules tell the robot what to do. For example, if the robot runs into a wall and cannot go forward it has been programmed to back up and turn left or right. Other rules might explain characteristics of the robot's universe—maybe the robot is in a small room. From these rules, the robot is able to learn from experience, adapt to changes, generalize about its world, and anticipate the future.

Another procedure under experiment is the Cerebellar Model Arithmetic Computer

TESTING FOR MACHINE INTELLIGENCE

Back in the 1950s, the British pioneer in artificial intelligence, Alan Turing, created a test for determining machine intelligence. The object is to have a person converse via keyboard with two subjects—a computer and another person—who are hidden from view. The person asks questions of the two subjects in order to decide which answer came from the computer and which from the human. If the person asking the questions is wrong and can't tell the difference at least 51% of the time, the computer has successfully simulated human intelligence.

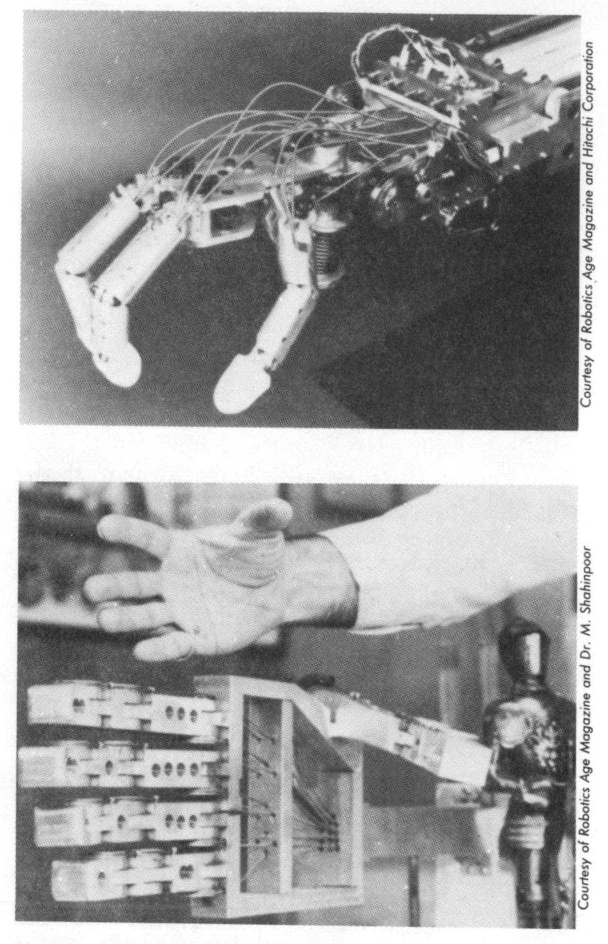

Courtesy of Robotics Age Magazine and Hitachi Corporation

Courtesy of Robotics Age Magazine and Dr. M. Shahinpoor

The next generation of grippers may look more like the human hand.

(CMAC), which is designed to work more like the human cerebellum (the part of the brain that controls the movement of arms, legs, eyes, and hands). The CMAC gives a command to the machine. The robot moves. That motion is monitored by sensors and sent back to the computer, where it is analyzed before the next command is given. A robot can actually monitor its progress and make corrections as needed. The system works reasonably well when the robot is making sweeping, continuous movements. It is less effective in controlling sporatic, discontinuous movements. The coordination between brain and body is a process we perform thousands of times every day without even thinking. Yet, for the computer or robot it is a very slow and clumsy process.

SIX LEGS WERE MADE FOR WALKING

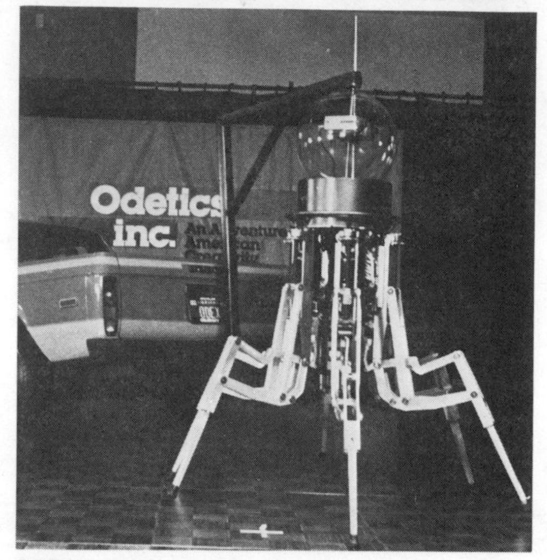

Courtesy of Odetics, Inc.

ODEX I is strong as well as agile.

Ever since you were a year old, walking has been one of the easier things you do each day. But if you think about it, the simple sequence of throwing one leg forward and landing on a foot involves innumerable sensor-motor messages. The balance-sensors in the inner ear communicate to the balance control center in the brain. The motor-control areas of the brain's cortex send messages to the muscles in the leg, foot, and back. Similarly, the nerves in the joints and muscles relay messages back to the brain—all this just so you can walk across a room. Oh, don't forget the role vision plays in walking! It's no wonder most mobile robots roll rather than walk upright.

In 1983, however, Odetics, Inc. unveiled the ODEX I, a six-legged functionoid. This robot has brought new meaning to the words robot motion. ODEX I can walk up and down stairs and maneuver around equally well in a coal mine or on a rooftop. With six legs, ODEX I can raise three legs for walking and still have the stability of a tripod. In addition to being versatile and stable, the functionoid is strong. Each leg can lift 400 pounds.

THE ROBOT WITH TV EYES

Given the clarity with which you see pictures on television, you might think that a television scanner would make a good robot eye. The problem is that a robot's on-board computer must digitize the camera's image into pixels and assign a number to each color or shade on the pixel grid. A robot does not see a television picture, only a pattern of numbers that represent a picture. The robot's computer then evaluates the data for recognizable shapes. If a shape matches with a previously recorded pattern, the robot will "recognize" the image.

This is a time-consuming process. And, if the on-board computer is also being used to control a robot's movements, speech, or voice-recognition system, the process will take even longer. Some robots take as much as 10-15 minutes to analyze an image. In that length of time, the sun will change the shadows on an object. The robot will see a new scene and have to analyze the image all over again.

Engineers are trying to develop methods that will help the computer recognize images more quickly. One way is to program the computer to build a whole image from fragments. This way a robot would not have to match an image pixel for pixel to recognize it.

WHAT'S THE DIFFERENCE BETWEEN A LIFESAVER AND A WASHER?

Can you tell the difference between a Lifesaver candy and a metal washer? Of course you can! Even if you can't see the two objects, you can pick them up and feel the difference. A Lifesaver is thicker than a washer. It is also a lot stickier. Put both objects in your mouth and you are certain to tell the difference.

A robot, on the other hand, might fail this simple test. With its primitive pixel-imaging vision, a Lifesaver and a washer look remarkably similar. Both are round with a hole in the middle. Some robots could pick up the objects and use a tactile feedback system to determine that the Lifesaver is thicker. But can a robot measure stickiness? What's more, without a mouth it can't taste the difference.

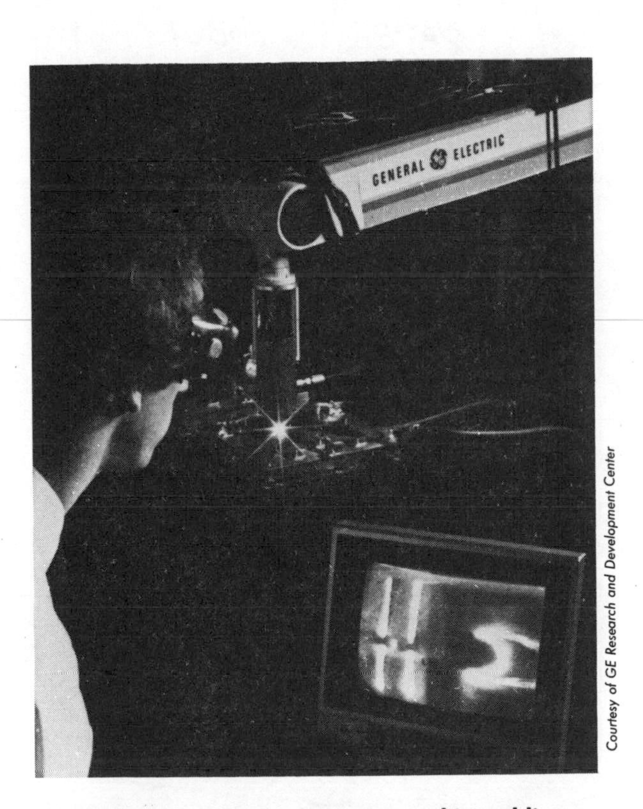

Courtesy of GE Research and Development Center

While no match for the human eye, this welding robot's vision system enables it to steer along an irregularly-shaped joint.

DIGITIZE is a process computers use to convert information (in this case images) into a numeric form that computers can understand, specifically zeros and ones.

PIXELS is short for picture elements. Look closely at a picture in a newspaper or on television and you'll see lots of little dots. Those are pixels. Computers assign a numeric value to each dot.

FEEDBACK is input data returned to the robot in response to some output. When your tastebuds tell you that a Lifesaver is peppermint-flavored, that information is feedback.

EVOLUTIONARY ADAPTIVE MACHINE INTELLIGENCE

Homebrew robot designer and author, David L. Heiserman, has developed the EAMI (Evolutionary Adaptive Machine Intelligence) Learning System to let computers and robots develop their own conclusions about their environment. EAMI offers three levels of learning.

At the *Alpha Level,* the machine operates on reflex. If a robot bangs into the wall, it has been programmed to turn and go in another direction.

Each time the robot bangs into the wall and turns, it stores the experience in memory. *Beta Level* learning involves taking those stored responses and rank ordering them according to success. If turning to the right before proceeding forward maneuvered a robot out of a corner more often than turning left, the robot would rate right turns as the better choice.

Finally, after banging into enough walls and maneuvering out of corners, the robot is ready for *Gamma Level* learning. That occurs when the robot begins to generalize about how to escape from tight corners. The robot should come to the conclusion that upon coming to a wall or other immovable object, turn right. Experience has told the robot that this is the way to avoid getting stuck in a corner.

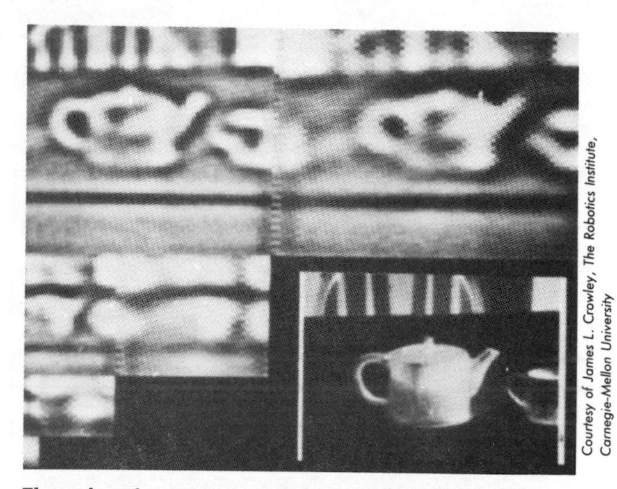

Courtesy of James L. Crowley, The Robotics Institute, Carnegie-Mellon University

These band pass images of a teapot exemplify an experimental shape matching process developed by James L. Crowley at Carnegie-Mellon University.

In 1984, it is safe to say that robots are no match for people. And yet, in spite of everything that robots cannot do, the promise for the future remains bright. Those involved in robotics are optimistic; they say that robots are inevitable. With more memory, more and faster microprocessors, and more sophisticated sensory systems, robots will be able to take over much of the physical labor in the workplace and in the home. On the other hand, scientists and researchers cannot agree on when this will happen. Educated guesses range anywhere from 10 to 30 years.

In the meantime, we will need to make some major changes. We will need to find

Illustration by Bob Johnson

new work opportunities that challenge human skills. People will also need new outlets for creative expression and more ways to spend their leisure time. Eventually we will even have to face such social issues as the possibility of robot-human marriages and the status of bionic people. How many parts of a human body can be replaced with bionic (machine) parts before a person ceases to be human?

Think about the future. How do you think you might be affected by robots? How closely do you think robots will resemble the Hollywood image? Will robots ever look and act like people? Whatever the outcome, you can expect to see dramatic changes in robots over the next twenty years or so.

HOW TO RECOGNIZE A ROBOT

In spite of all the features that we have come to associate with Hollywood robots, the following checklist is all you need to identify a real robot:

1. Robots are versatile by design and can be programmed for different kinds of work;
2. Robots have feedback systems and can monitor their own progress using their on-board sensors;
3. Industrial robots represent a direct replacement for human labor (eventually domestic robots will too);
4. Robots have computers for brains;
5. Some, but not all, robots are mobile.

ROBOTS IN THE REAL WORLD

INDUSTRIAL ROBOTS

The largest and most important class of robots in the world today looks nothing like people. The industrial robot or robot arm is a sophisticated tool holder or "gripper" attached to a wrist and arm mechanism. This unit can be mounted on a stationary base, a wall, or even on wheels.

Because these robots are reprogrammable, they do not become obsolete as quickly as other machinery. To understand the advantages, let's look at the automobile industry. Car designs change every year, and conventional machinery built to perform a particular job on a particular car model must be discarded or retooled, at considerable expense, before it can work on a different model. A robot, on the other hand, needs only to be given new instructions (reprogrammed) and it will be ready to work on another car. This means big savings to the manufacturer.

Although industrial robots do not look human, they are direct replacements for human labor. In industry, some jobs are very monotonous. People become bored and inattentive in these jobs and are apt to make mistakes. Machines never get bored. They will do the same job the same way until someone stops them. Why not use machines for these jobs and hire people for more interesting and complex tasks? Similarly, other

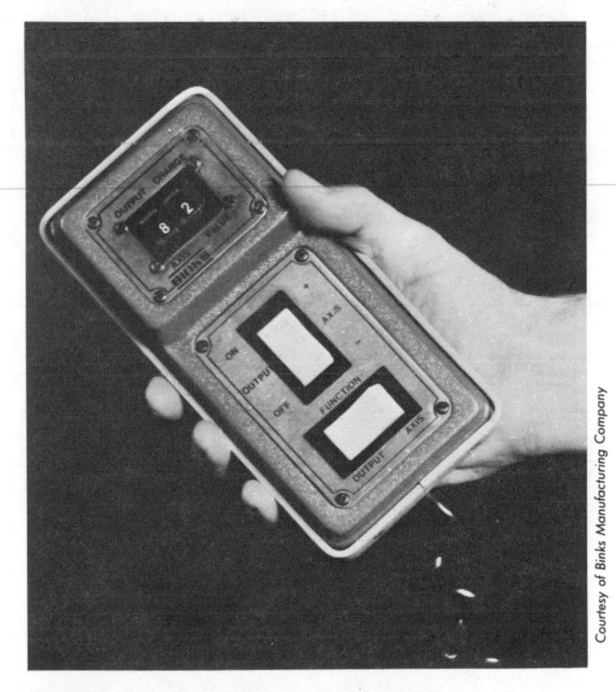

Courtesy of Binks Manufacturing Company

Many industrial robots are programmed with a joystick or handheld keypad similar to this one.

A NEW BREED OF COP

Felix and Snoopy are remote-controlled robots used to help police in dangerous situations. Because they are not flesh and blood, they can be sent in to photograph suspects "holed-up" in a building; they can defuse bombs and shoot tear gas. Alas, they can also die! Snoopy, used by the Oakland, California, police malfunctioned and came to a roaring stop in the middle of a street in clear range of the suspect's position. The police were unable to rescue Snoopy until the seige ended.

jobs, such as mixing poisonous gases, handling radioactive material, or working around hot furnaces, are dangerous. An accident could kill a person. Why not use a robot? An injured robot can be repaired or replaced.

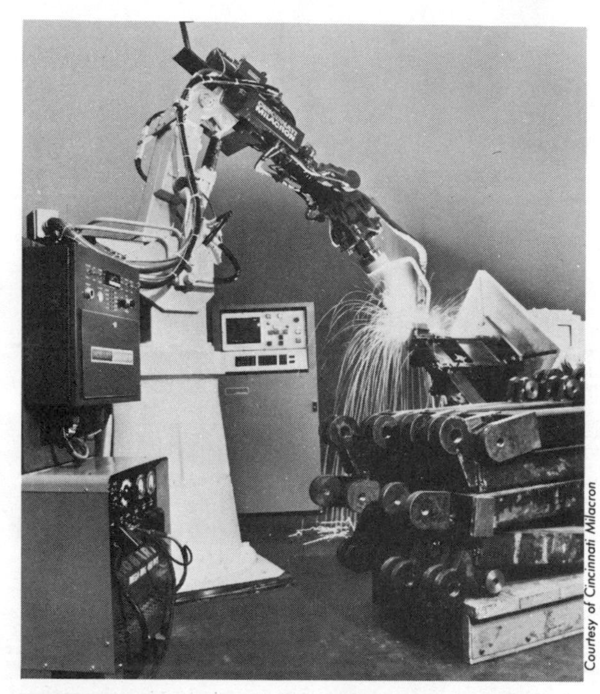

Courtesy of Cincinnati Milacron

Once programmed, the robot can continue this same welding task indefinitely. It never tires or gets bored.

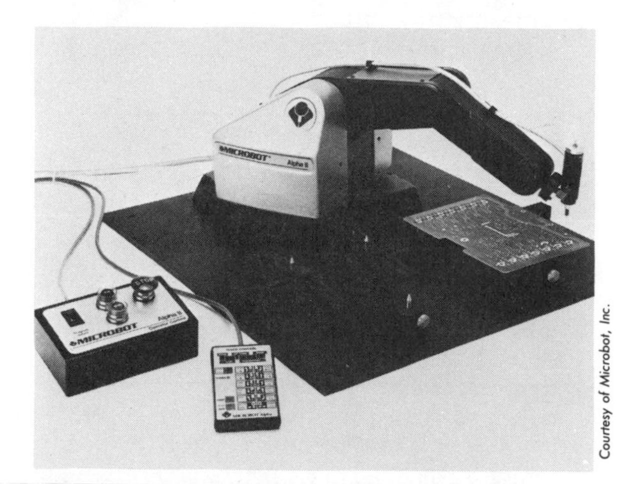

Courtesy of Microbot, Inc.

Industrial robots don't have to be large. This Microbot Alpha II is delicately etching a pattern on a printed circuit board.

Typically, you will find industrial robots doing welding or finishing work. When equipped with lasers, they can perform very precise operations. Finishing work includes sanding, grinding, and painting. Robots in automobile factories are often used to weld and paint new cars. Although vision is one area that needs further refinement, a robot with a camera-lens eye can inspect finished products moving down an assembly line. Using a gripper, the robot can even assemble small electronic equipment.

These are the conventional applications, but as the technology continues to evolve, industrial robots will perform more varied and difficult tasks. There really is no limit to the shapes of or applications for robots. Jason, for example, is a self-propelled robot with manipulator arms and stereo, color-television eyes. It is being designed to collect marine samples and map the ocean floor. Unlike a human diver who can only stay down for a few hours, Jason will dive for about a month before surfacing. The controlling equipment is stored in a twelve-foot, unmanned, sea sled that is tied to Jason by an umbilical cord.

The Japanese have been faster to accept robots than any other country. More than

CAN YOU SWALLOW THIS ONE?

The Japanese are experimenting with a robot model that resembles a snake. A very small, snake-like robot would be able to enter the human body, carrying a small television camera with it, and photograph a person's stomach. This could be very useful to doctors who are trying to diagnose some stomach or intestinal disorder.

BLINKY DELIVERS THE MAIL

Blinky is a mail-delivering robot in Washington's Department of the Interior building. It is programmed to follow a particular course and stop outside office doors. Because Blinky does not have arms, people must actually remove their mail from the slots on Blinky's sides.

With Calma's ROBOT-SIM software, engineers can simulate an industrial robot's movements on a computer screen before putting it into operation at the factory.

20,000 industrial robots are currently employed in Japan, while only somewhere between 5,000 and 8,000 are used in the United States. American manufacturers have been slow to adopt the new technology because the initial expense is high. Robots cost between $50,000 and $200,000 a piece, and most companies require more than one machine. But as companies are discovering, robots quickly pay for themselves. A company that manufacturers computer disk drives purchased a robot to do assembly work. The robot costs $150,000, but it replaced three people on the assembly line and paid for itself in only 18 months.

It's hard to argue with statistics like this, unless you are one of the people put out of work. Clearly, we have a potential problem here; one that requires careful thought and planning. Estimates vary from four million to a couple hundred thousand people who will lose their jobs as a result of robots. The question is, will enough of the displaced workers be rehired in other capacities? Can they be retrained? Can we begin now to prepare young people (our future work force) for a world in which they will be competing, and working, with robots?

Illustration by Bob Johnson

In fact, labor is already worried. Recently, a union charged the U.S. Labor Department of not informing them before a robot was installed in a government laboratory. Despite the fact that the robot relieved four men from a very dirty job of testing dust samples from mines, the union argued that the workers were given little consideration. The employees were not fired; they were given some "busywork" that union officials felt would soon end, at which time the workers would be out of jobs. At this writing, the union wants either the robot fired or the employees retrained or reassigned to some permanent work.

Many colleges and vocational schools around the country are adding robot design and repair courses to their curriculum. Educational robots, which are scaled-down models of industrial manipulators, are being used in these programs. Students receive the hands-on training they need to assure a good job upon graduation.

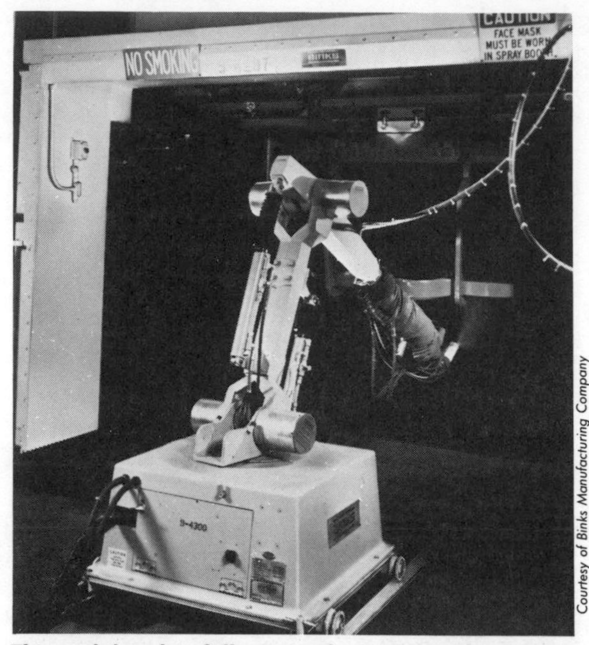

Courtesy of Binks Manufacturing Company

This mobile robot follows tracks to get to the spray-painting booth.

WALDO IN OUTER SPACE

Waldo is a fifty-foot giant of a robot arm used to move large objects on board NASA's Space Shuttle. It is named after Waldo J. Jones, the physically-disabled inventor in Robert Heinlein's science-fiction classic, *Waldo.* Waldo Jones suffered from a disease that weakened his muscles, denying him control over his arms and legs. He overcame the disability by creating a gravity-free house that orbited Earth and a pair of robot arms. Floating free in space, he controlled the robot manipulators with gloves that sensed his hand movements and duplicated the slightest motion. The Waldo on board the Shuttle is not designed for precision hand work. Rather than a gripper hand, it has only a wire snare and is controlled by joysticks.

RUN, ROBOT, RUN

Several tracking methods are used to guide robots. The simpliest is to put a robot on rails, like a train. Some factory owners avoid having tracks running around the factory by installing a cable under the floor. A robot is then fitted with two electric-coil sensors that detect a magnetic field when current flows through the cable. An on-board computer keeps an equal charge flowing through the two coils by guiding the robot along the path of the cable.

Other robots follow florescent lines painted on the floor. Invisible in normal light, the line only shows up under ultraviolet light. Two light sensors are positioned under the robot. The distance between the sensors is slightly less than the width of the painted line. They are adjusted to the amount of light reflected off the line. If the robot veers off to the right, the right sensor will be the first to detect the change in light. It will send a message to an on-board computer or microprocessor to steer the robot back toward the left. If it goes too far, the left sensor will direct the robot to steer a bit to the right.

HOMEBREW ROBOTICS

If you enjoy tinkering in electronics and mechanics and can imagine controlling a Big Trak toy tank with a hand-held computer, then you may be one of those special people we call homebrew robot builders. Connecting computers, motors, sensors, and mechanical parts is the heart of the homebrew robot builder's work.

While homebrew robot builders tinker with computers and electronics for fun, they are actually vital to the robot industry. Although their devices are not designed to increase manufacturers' productivity, homebrewers often create impressive, sometimes bizarre, robots out of a handful of spare parts. They enjoy experimenting on the fringes of technology. Homebrew builders may spend all their time playing around with radar sensors or robot grippers. In the process, if they should stumble on a particularly good design for, say, robot motion or a feedback system, the whole industry benefits.

All their influence, however, is not reserved for the industry. Homebrew builders touch the home market as well. Their effect is similar to that which computer hackers have had on the personal-computer revolution. Just as hackers have shown us what we can do with a home computer, homebrewers will probably give us practical ways to use personal robots.

Homebrew robotics began in the period after World War II. As military stockpiles turned into available surplus, mechanical and electrical equipment and spare parts could be bought cheaply. Soldiers who went back to school were often exposed to electronics and mechanical engineering. Some even learned about new theories in communication and automatic control systems (known as cybernetics).

In the 1950s, the new computers spark-

Photograph by Chad Chadwick

Gene Oldfield at Robot Repair, his homebrew workshop.

ed the imagination of many a garage homebrewer, but the systems were too big and too expensive. Computers, which in those days could fill rooms, were not available to or practical for the home tinkerer. Homebrewers had to be content with "relay rats" and "tubeturtles" built from surplus electrical relays and vacuum tubes from old radar equipment. But "rats" could only respond to a few things and their ability to "think" was limited at best.

When transistors were introduced for commercial use in the 1950s and the transistor radio was produced, experimenters were excited. It was just the beginning of miniaturization. The 1960s saw the development of the integrated circuit and by 1967 the TTL chips (which stands for transistor-transistor logic) became an industry standard for digital electronics.

Thanks to the new technology, the possibilities for robots became more real. They wouldn't rival the Hollywood image, but they were a start. For example, minicomputers were linked with mobile platforms. These mobile computers were used to study

Courtesy of SRI International

some of the basic problems in robotics, such as controlled movement within a two-dimensional plane. These were important first steps.

A small revolution in electronics secured the future of robotics. With the creation of the microprocessor chip, it was possible to lay down thousands of wires in an instant and program the circuit later. In 1976 MOS Technologies introduced the KIM, a single-board computer with hexadecimal keypad (a numeric keypad for entering machine language programs) and an LED (light-emitting diode) display. The display panel served as a small monitor. The KIM also had input/output (I/O) ports, allowing the computer to control external devices. This little engineering computer also featured the 6502 chip,

LED stands for light-emitting diode, which is a bead of colored plastic that lights up when a battery is attached to its two metal leads.

A SHAKEY EXPERIMENT

In the late 1960s, Stanford Research Institute International began experimenting with a mobile, camera-guided robot called Shakey. While somewhat unreliable (hence its name), Shakey could "see" three-dimensional objects and navigate around them. It even demonstrated a primitive form of artificial intelligence. Shakey didn't just roll about a room, it planned its movements. By saving to memory the results of each movement, Shakey learned and even made generalizations based on these experiences. Shakey as it was, it proved that robots could be built with the technology of the day!

Shakey was upgraded in 1971, using more sophisticated hardware and software. It was programmed to take blocks from room to room and assemble them in a particular fashion, as well as to open doors and roll up and down a ramp.

ZOOMING IN ON SONAR

For those who can't find a broken Polaroid and don't want to take apart the family camera, Polaroid has produced the Ultrasonic Ranging System Designer's Kit. It includes a technical manual, application notes, Ultrasonic Transducer, circuit board, LED display, two Polaroid Polapulse 6VDC planar batteries, wires, and connectors. The kit costs $165, which is a bit steep, but you might still want to request some information for your files. Write to Polaroid Corporation Commercial Battery Division, 784 Memorial Drive, Cambridge, MA 02139.

which is found in Apple, Atari, and some Commodore computers.

By the late seventies, experimenters had everything necessary to create a robot—homebrew style! Large, motorized toys with gear-driven wheels; sensors; and computers were about all that anyone needed. A homebrew builder merely had to scrounge the parts, build a frame, interface the electronics and mechanics, and add the sensors.

New developments have continued to put better and less expensive technology within the reach of the homebrew builder. For example, when sonar was first developed, it was expensive. The government could afford to use it in submarines, but a robot builder could not even consider sonar. A few years ago, however, Polaroid introduced the Ultrasonic Ranging System for its cameras. Homebrew builders discovered an inexpensive source for sonar sensors and began scrounging around for broken Polaroids.

The growing portability of personal computers, starting with the Sinclair, has been a boon. Many homebrewers are now looking at hand-held and briefcase computers. Do not be surprised to find someone interfacing the I/O ports to a couple of motors to create a cheap, light, robot controller that responds to programs written in FORTH or Logo. Because of the advantages of mobility, the robot is a natural extension for the personal computer.

Photograph by Chad Chadwick

You'll find everything from the yoke of a little red wagon to nonvolatile memory on board Entropy.

ENTROPY—BUILDING A ROBOT FROM SCRATCH

Gene Oldfield, began building his first major homebrew robot around 1976. Entropy, as it was called, was a mobile, three-wheeled robot powered by a car battery. A KIM single-board computer was interfaced to the sensors and relays by only seven microchips, which means that most of the processing was done in the computer itself. To give you some idea of the process a homebrewer goes through, we have outlined the steps in Entropy's development:

"I began," says Gene, "with the mechanical construction. Entropy's back wheels were on a common axle, which means that the turning center must lie somewhere on the extended axle. Rather than build one from scratch, I used the axle yoke and wheels from a discarded toy red wagon." The front wheel was motorized and attached by a vertical axle (called the scan). The electrical connections to the motorized wheel were commutated using generator brushes from a car motor. "Had I wired the wheel and motor instead, the wires would have become twisted and eventually broken from all the turning required to steer the robot." A second motor with gearhead drove a gear on the scan axle along with a cam assembly that allowed the robot to be set in any direction.

"I made the frame of wood," Gene continues, "and painted, waxed, and covered it with copper foil." While not a traditional material for building robots, wood has advantages. It's an electrical insulator; light, strong, easily worked; and it's very simple to attach switches, wires, and terminal blocks as you go along. Also, wood is not as high tech and makes the robot acceptable, more like furniture. The visual impact is an an important design criterion. A robot should be friendly, fun, and nonthreatening. "As a homebrewer, you can get away with a lot and still make your creations legitimate. R2D2's popularity was due, in large part, to being cute."

Gene decided to wire Entropy on a protoboard (also called a breadboard). 'Since I was developing Entropy for the first time," he explains, "I did not know how it was going to be wired." The protoboard permits endless experimentation. Once power wires were in place, it took about four hours to connect Entropy's relays using patch cords (wires). "This is both fun and satisfying—you feel that you could emulate anyone's robot by simply changing a few wires." At that point, Entropy was in the same category as the mechanical/electrical rats and turtles of the fifties. It could center its scan axle and stop.

Next, Gene wired the motors, sensors, and relays to the electronic components that interfaced with the on-board computer. Again, because the pattern of connection was not established, he used a protoboard.

Entropy was to operate without external wires, hook-ups, or data links. Data from sensors was processed on board. While doing the dishes and other seemingly simple tasks were impossible, Entropy did have the sensors and memory to be effectively mobile. It could travel from room to room and navigate through doorways. Sonar, using a tone-decoder chip, measured multiple reflections on distances ranging from one inch up to fifteen feet.

The KIM computer on board allowed

Illustration by Bob Johnson

ROBERT PROFETA—HIS ROBOTS ARE ROLLING ELECTRONICS PROJECTS

"A good robot builder," says 21-year-old Robert Profeta, "has to have the right kinds of interests. It helps if you like taking things apart to see how they work." As a boy, Robert was always taking apart toys, clocks, and small appliances. He enjoyed chemistry, but preferred electronics. "A person can really experiment with electricity and electronics. You don't have to understand how everything works before you start. I learned as I went along." As long as you are working with small, DC currents, your experiments are not terribly dangerous—even if something goes wrong.

Robert built his first robot for a ninth-grade science project. It was mobile, had an electronic eye, touch sensors, arms, and LEDs for running lights. His second project was a two and a half-foot scale model of R2D2. "R2P2 (after Robert's initials) started out as a clear, plastic terrarium and a lard can," says Robert. When finished, R2P2 could move across the floor and turn its head. While R2P2 has no practical functions, it looks nice.

His third and fourth projects were both visually appealing and technically interesting. Robert thinks both elements are important. Rover and DC Prober are mobile and include such features as a photoelectric burglar alarm, heat sensor, sound sensor, infrared detector, soil tester, and water-purity tester. "I included so many sensory devices because I wanted to demonstrate robots' potential to be functional. It's a prototype with examples of what is possible."

Robert suggests that robot builders avoid a lot of frustration by starting small. Also, build a lot of electronic kits and incorporate some of the projects into your robot designs. Read books and magazines to get ideas for new projects. "Above all, keep experimenting—it's the best way to learn."

Courtesy of Robert Profeta

Robert Profeta's robot Rover

JONATHAN KAPLAN—HIS HOBBY HAS BECOME A CAREER

Jonathan Kaplan traces his interest in robots back to when he was four and watched cartoons about robots. He built his first robot when he was eight. Today, Jonathan is a graduate student, studying artificial intelligence with the Robotics Research Group at the Massachusetts Institute of Technology.

All total, he has built around a dozen robots, not all of which are still working. "I keep the complex ones," he explains, "others get recycled for the parts." He also scrounges for parts in electronics supply stores and junk shops.

Unlike many homebrew builders, Jonathan does not name his creations. He also does not treat them like pets. As he says, "I am painfully aware that they are only machines." He feels, however, that they can be smart machines! He even built an intelligent robot with sensors to search out and avoid objects. The robot stands about 33 inches tall. His ideas for this and most projects come from reading books and magazines . . . and a lot of trial and error.

When beginning a project, Jonathan starts with the problem itself—what he wants the robot to do. He is usually uncertain what elements will go into producing the desired result. Once the problem is outlined, he breaks it down into a series of smaller processes until he is left with problems he can solve. He then tackles the individual projects, using any device that meets his needs. He solves the control aspects (electronics) first because they are more abstract problems and involve more creativity. "If you don't understand the theory and control aspects," he warns, "the mechanical part will never come together."

Jonathan did not have a lot of help on his projects. He encourages other robot builders "to explore ideas and pursue them, even if there is no one around to help you, and by all means follow your own imagination."

Courtesy of The Robot Exhibit at the American Craft Museum, American Craft Council, New York City, 1984. Photography by Ralph Gabriner.

Jonathan Kaplan created the Intelligent Robot between 1977 and 1979. This hobbyist robot is mobile and programmable and has dimensions of 14″ x 14″ x 36″.

Photograph by Jeff Lundell

Gene turned this motorized toy car into a robot with the aid of a few switches, an automobile windshield wiper motor, and a single board computer.

10 HINTS FOR HOMEBREW ROBOT BUILDERS

1. Build first, dream second. You do not need a computer to start building the frame, wheels, and motor sections. Get the skeleton in working order; the computer and feedback sensors can be added later.

2. Plan for ignorance. Unless you have built a particular robot before, you probably do not know exactly how the wires should be hooked up. Experimenting is easier if wires are attached to a terminal strip or protoboard (also called a breadboard). Changing and improving your designs will be easier.

3. Keep it simple. Be realistic about the problem. The robot will be complicated enough to build without trying to build a revolutionary maneuvering system.

4. Consider modifying a large motorized toy. Many good homebrewers have started this way.

5. Keep your robot light. Consider weight when choosing parts.

6. Keep a notebook of everything you do. You are going to have to repair your creation someday, or you may decide to modify it. Without good technical notes you will have trouble remembering just what you did.

7. Plan to work alone. Robot creation by committee is difficult.

8. Consider using a KIM or SYM single-board computer with built-in hexpad and display. You will have to program in machine language, but you'll learn. Install a 6116 CMOS (Complementary-Metal-Oxide-Semiconductor) nonvolatile RAM chip so that your basic robot programs are held in memory.

9. Avoid complicated interfacing. Rely on the computer. It already has the capacity to do most of the processing you will need. You should not need more than one or two chips to interface the computer to the robot.

10. Make the feedback come from the room. A sonar device that bounces sound waves off the furniture or a vision system that allows the robot to see objects are good examples.

Entropy to be programmed for motion at over 100 instructions per minute. Operations were performed through the record/playback program, much the way you save, load, and run programs on your computer. In the playback mode, the computer sends out commands, which the robot acts upon.

Since Entropy lacked a tape recorder or disk drive, Gene used nonvolatile RAM. With a special CMOS RAM chip (which requires very little electricity) and a couple of nicad batteries, he could keep the RAM powered up even when the rest of the system was turned off. The system is both cheap and reliable. Programs are always present. Simply turn on the robot, select a program's starting address in memory, and press go.

Entropy was a successful project. The robot wandered around the house, moving from room to room.

ROBOTS COMES HOME

In September 1983, an RB5X personal robot in Columbus, Ohio, placed a call to another RB5X in Denver, Colorado. The second robot answered the phone and listened as it's Ohio counterpart commanded it to dance. The Denver robot was then supposed to do a little dance and report back to the one in Ohio. "It would have worked," says Sharon Smith of RB Robot Corporation, "except that the phone connection was bad and the robots could not hear one another well enough."

Still, this first transcontinental call between robots suggests that the recent robot phenomenon has not been confined to the realm of the homebrewer's garage or the automotive assembly line. In fact, the greatest hope for the future is what robots will be able to do around the house. Personal robots, both assembled and in kits, have made their way into the marketplace. They are being bought by "the man and woman who have everything else" and by the experimenter who may not want to scavenge Silicon Valley garbage pails for spare parts or frequent flea markets.

The year 1983 was unofficially dubbed the year of the robot. The stars were three personal robots named HERO 1, Topo, and RB5X that range in price from $1500 to $2500. Other robots include Hubot, RoPed, Droid-Bug, Robocycle, and X-1. Currently, the distinguishing feature in personal robots is whether or not they have a computer (brains) on board.

To the less experienced technophile, these home robots represent little more than high-tech pets that are fun to have around the house. Their arms are not strong enough to do manual labor, they lack the reasoning capability needed to prepare dinner, and their

NICAD is short for nickel cadmium. While nicad batteries are more expensive than regular alkaline batteries, they can be recharged.

Illustration by Bob Johnson

sensors are not refined to where they can clean house without banging into the furniture. Unlike factory work, few household chores can be routinized. For example, try to write a program that would direct a robot to clean your room. Still, robots can be programmed to sing, dance, tell jokes, and serve food and drinks at parties. They can also be programmed to help with homework, drilling students on vocabulary words or giving them math problems to solve.

To the experimenter, however, even these early models, unsophisticated as they are, represent a challenge. They are interested in developing better languages for communication, more accurate voice recognition systems, and software that outfits the robot to do something more useful than wait on guests. One of the greatest challenges is the development of a motion algorithm that makes it possible for the robot to map and locate itself. Once robots can maneuver themselves about, the possibilities grow enormously.

Expect the commercial, personal robot companies and individuals to form users' groups that will make software available and promote unified development efforts. Furthermore, these companies will continue to push the technology to its limit.

As a robot builder, you will learn a lot by studying data on the design and capabilities of personal robots. Consult the following list of personal robot manufacturers. We suggest that you write to some of the companies and request their information packages. Then, start a file of information on robots.

Courtesy of RB Robot Company

Today RB5X just fetches the paper. Someday it may read the paper.

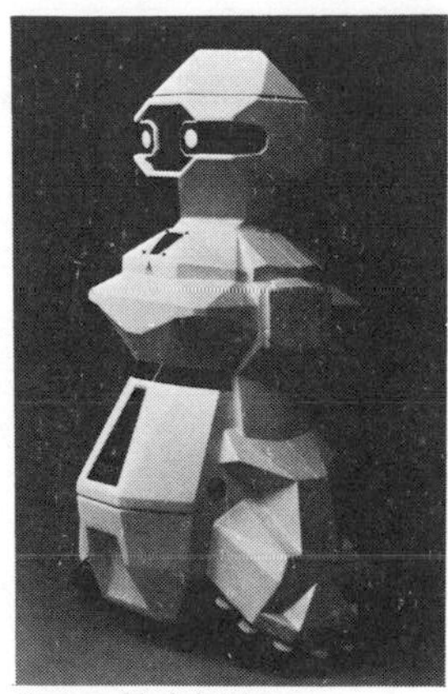
Courtesy of Androbot, Inc.

Digital Topo ($1500) is mobile, has speech capabilities, two-way infrared communication, and a card cage for expandibility. **B.O.B.** ($2500) has a computer on board. It also has some artificial intelligence capabilities.

Androbot
101 E. Daggett Dr.
San Jose, CA 95134

Colne Robotics
207 NE 33rd St.
Ft. Lauderdale, FL 33334

Armdroid 1 ($1295-$1600) has an on-board computer that controls 5 motors for arm and gripper; movement can be programmed through a remote computer.

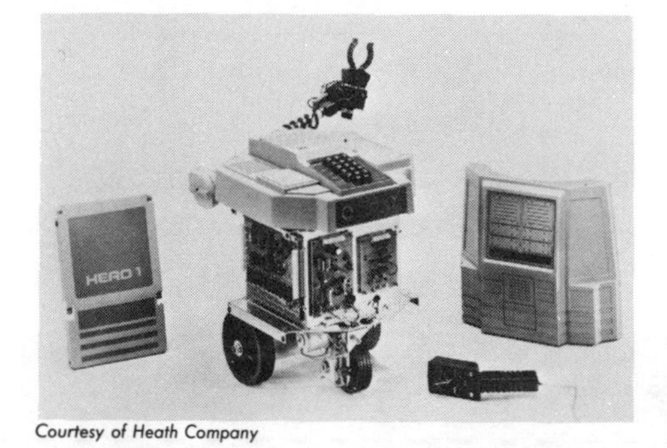

Courtesy of Heath Company

HERO 1 ($1500 kit & $2500 assembled) has a 6808 microprocessor on board. Program through a hexpad. HERO is mobile, head and arm rotate. The gripper can lift 1 pound. It has speech capabilities and voice recognition. Heath also sells a series of Movit kits ($40-$79) that feature various sensory systems, such as sound and infrared sensors.

Heath Company
Dept. 529-138
Benton Harbor, MI 49022

Hobby Robot Co., Inc.
P.O. Box 887
Hazlehurst, GA 31539

Smart Rabbit Robot kits ($400-$800) are compatible with several brands of personal computers. The robot is 22'' tall and weighs 25 pounds. It has a humanoid shape; arms move on more expensive models.

Courtesy of Hubotics

Hubot ($3495) is a home entertainment center with Atari VCS, AM/FM radio, cassette player, TV, clock, 128K computer, and serving tray. It will respond to voice command, sound fire alarm, and soon will include a vacuum attachment.

Hubotics
6352 Corte de Abeto,
Suite D
Carlsbad, CA 92008

Iowa Precision
Robotics, Ltd.
908 10th St.
Milford, IA 51351

Marvin Mark I ($6000) is a mobile, educational robot with voice, sonar vision, moving arms and head. It has an on-board 68000 computer (128K-512K) and floppy drive. A six-slot card cage is concealed in the humanoid body.

RB Robot Corp.
18301 W. 10th Ave.
Suite 310
Golden, CO 80401

RB5X ($1495 base price) is designed to move forward with technology. It has lots of expansion ports. It currently has voice recognition and speech synthesis, sonar, touch-sensitive bumpers, plain-language control system, and an arm. New features, including ROM-cartridge software, are added regularly.

Courtesy of Personal Robotics Corporation

RoPet (XR-$2199 & HR-$2499) has a Z80 processor on board to operate the collision avoidance system, voice recognition, and speech synthesizer. The systems are expandable through plug-in ROM cartridges. The basic difference between the two models is that HR comes as a finished robot, while XR includes just the bare workings. An experimenter could design his/her own shell around the XR system.

Personal Robotics Corp.
469 Waskow Dr.
San Jose, CA 95123

Courtesy of Rhino Robots

Rhino XL series ($2850 base price) interfaces to an Apple through RS-232 port. This 32" robot arm can lift 5 pounds. **Scorpion** kit ($700) is a mobile platform with on-board processor, sound generator, vision, bumper sensors, RS-232 interface, and 6VDC motors.

Rhino Robots Inc.
P.O. Box 4010
2505 S. Neil St.
Champaign, IL 61820

Photograph by Jeff Lundell

Robocycle ($995-2495) is designed with education and experimentation in mind. This mobile robot speeds along at up to 10 mph. The on-board computer has nonvolatile memory, an assembler, and RS 232 connection for developing programs using a remote computer like the Commodore 64. It has 50 to 70 I/O lines.

Robot Repair
816 1/2 21st St.
Sacramento, CA 95814

Courtesy of the Robot Shop

Droid-Bug kit ($100) has an obstacle sensor that causes the robot to buzz and turn away from objects. **X-1** kit ($300) has on-board computer control, solar battery charger, remote radio control, sound and obstacle sensors. A start-up package with pictures, data, and catalog is available for $5, which is refundable with first order. You can also buy individual parts to use in your own projects.

Robot Shack
P.O. Box 582
El Toro, CA 92630

BASIC ELECTRONICS—THE ELEMENTS OF ROBOT DESIGN

While we cannot see electricity, we know that it exists. In fact, these days we really take it for granted. We have proof every time we flip a light switch or turn on the stereo. While scientists have never actually *seen* electricity either, they have measured electrons and documented the many properties of electricity. For example, they have determined that it flows through wires and other conductive materials much the way water flows through a pipe. In fact, that analogy may help you to understand some of the principles of basic electricity and electronics.

While the use of electricity to operate the many appliances that we take for granted is relatively recent, people have known about this wonderful force for a long time. As early as 640 B.C., Thales of Milago, a Greek scientist, observed the properties of amber (a yellowish-brown, sap-like resin that resembles plastic). When rubbed briskly against fur or cloth, the amber attracts small, light-weight objects like paper, cloth, and wood chips—almost like a magnet. The Greeks thought that this was a special power in the amber and called amber "elektron." In time, people came to understand this phenomenon as *static electricity.*

In 1791, an Italian surgeon, Luigi Galvani, noticed that the legs on a dead frog he was dissecting moved whenever he touched them with his metal dissecting instruments. He thought that electricity was trapped inside the frog!

Just a few years later, another Italian, Allesandro Volta, figured out that zinc and copper dissecting tools (two different metals) and a strong, corrosive liquid (the frog's bodily fluids) created current electricity. The knowledge led him to create the first battery. This was an important step since the battery was the first source of continuous electrical current. Static electricity is not constant enough to be a power source.

Experimentation continued throughout the nineteenth century. A Danish scientist, Hans Christian Oersted, actually documented the direct relationship between electricity and magnetism. You can test this yourself: Take a large metal bolt or spike, a sufficient length of enamel-insulated bell wire (also called coil wire), and a 9-volt battery. Carefully wrap the wire around the bolt or spike 50 times—taking care to make a neat coil. Attach the battery to the two ends of wire. Remove just enough of the insulation to get a good connection. Notice that when the leads are connected, the spike makes a good magnet, an electromagnet to be exact. *Handle carefully,* the magnet may get hot.

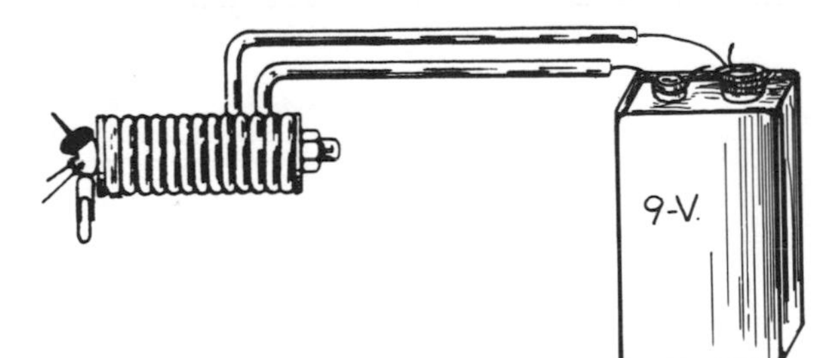

Similarly, in the 1830s, Michael Faraday demonstrated that a magnetic force (created by moving a magnet through a wire coil) could be converted into electrical energy. This finding is the basis for the DC motors that you will use to operate your robots.

As scientists continued to experiment, the various basic principles and properties of electricity were discovered and named. Mastery of these basics is vital to your understanding of the electronics needed in robotics technology. So, if you are unfamiliar with any of the following, take the time to learn the terminology and the principles involved. If you feel that you need further explanation, you might want to consult a basic electricity or electronics handbook (see the list at the back of the book for suggested titles). On the other hand, if you are comfortable with this material, move on to the robot project on page 45.

The flow of electricity through wires is referred to, simply, as electric CURRENT. Stated more scientifically, current is the movement of charged particles in a conductor. Current is described as being either *direct* (DC), which comes from a battery source and flows continuously in only one direction, or *alternating* (ac). Household electricity is alternating current. For the purpose of our robotics experiments, we will only need to concern ourselves with direct current. All electrical current is measured in AMPERES, which refers to the *amount* of current flowing through a circuit.

The strength of a current is determined by the pressure that is applied to a wire. This pressure is measured in VOLTS. The water pipe analogy may be useful here. The pressure as water flows determines the force of the water. The greater the pressure, the stronger the water's force and the faster it flows. You can actually see the difference. The same is true of electricity. The greater the current's pressure against a wire, the higher the voltage and the stronger the electrical force. The amount of water or electricity flowing does not change, just the force pushing it through a pipe or wire.

REMEMBERING OHM'S LAW

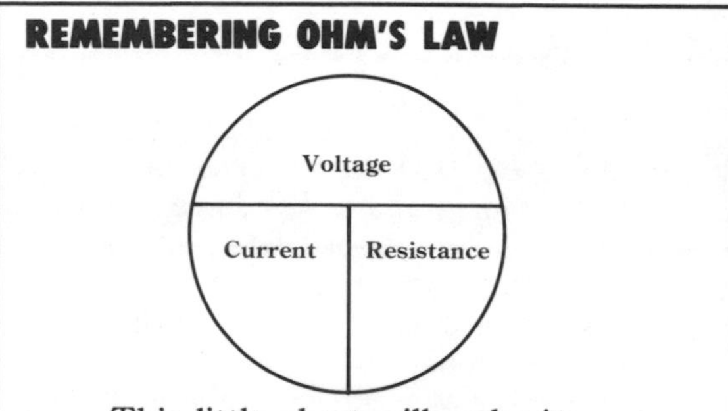

This little chart will make it easy for you to remember Ohm's Law. Memorize the chart, and you will have all three equations at your finger tips. If you want to know the equation for voltage, put your thumb over VOLTAGE and read what's left. The CURRENT is next to the RESISTANCE. Therefore, **Voltage** = **Current** x **Resistance.** If you want to know the current, put your thumb over CURRENT. You will see VOLTAGE over RESISTANCE. Therefore, **Current** = **Voltage** / **Resistance.** Finally, by putting your thumb over the RESISTANCE you can figure out that **Resistance** = **Voltage** / **Current.**

The water streaming from the tub on the right is under greater pressure (voltage), but has less volume (amperage).

The water streaming from the tub on the left has greater volume (amperage) and less pressure (voltage).

HOW DOES CURRENT FLOW?

There is considerable discussion these days as to whether electricity is the flow of holes created by the absence of electrons (positive charge) or the flow of electrons (negative charge). For all practical purposes, the result is the same—charged particles (either negative or positive) are attracted to particles with the opposite charge. The only difference is that electrons flow in the opposite direction of the holes. In this book, we will stick to the theory of Conventional Current Flow in which a positive charge flows to negative. Just be aware that you are likely to come across experiments elsewhere that describe electricity as moving from the negative power source to the positive.

Unlike water, however, current needs more than voltage to flow. It has to have someplace to go! It requires a path, or CIRCUIT. No matter how complex a circuit, the electricity always flows from the power source (e.g., battery or wall socket), through the wires, electronic components, or other conductive materials, and back to the source. You made a circuit when you wrapped wire around the bolt and attached the two ends to a battery. The design of a circuit determines how *much* current will pass through a conductive material before overheating. This degree of conduction is called RESISTANCE and is expressed in OHMS. Returning to our water analogy, if you step on a hose, decreasing the diameter of the hose, you have created resistance. You are making it harder for the water to flow. Pressure builds up inside the hose. If you could restrict the flow

#1 USING OHM'S LAW

Let's say you want to attach a small LED to your robot's electrical circuit. The LED requires only 20 ma (milliamperes, 1000 ma = 1 ampere) from a 9-volt battery. To keep from burning out the diode, you will need a resistor to limit the amount of current. The schematic would look like this:

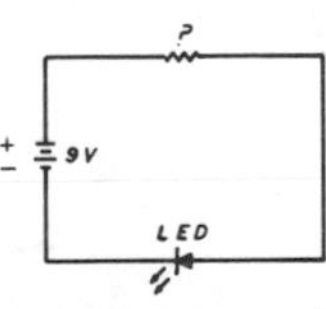

Using Ohm's Law, in the form: **Resistance (R) = Voltage (E) / Amperes (I),** you can figure out the size resistor needed. Convert milliamperes into amperes (20 ma = .02 amperes). **R = 9VDC / .02a;** the resistance equals 450. You need a 450 ohm resistor. *If you cannot find a resistor with exactly the number of ohms you need, use the next largest size.* In this example, you will probably have to use a 470 ohm resistor. (Probe around the LED with your voltmeter. You will see a .7-volt drop, called a *junction drop,* across the component. In most silicon semiconductors (e.g., transistors and regular diodes), there is a .6 - .7-volt drop; with germanium, there is a .3 - .4-volt drop. While junction drops may not affect your current work, be aware that they exist. As you get into more complex projects, requiring greater precision, small drops in voltage may throw off all your calculations!)

In selecting the proper resistor, you also need to know the wattage. A **WATT** is the unit of electrical power represented by the amperage times the voltage of a circuit. In other words, some amount of energy is used when pressure is applied to current to make it flow through a circuit. Similarly, it takes energy to create the force to push water through a pipe. You can determine what that amount is just by multiplying voltage by amperes. Figure out the wattage of the resistor in your LED circuit: **Watt = 9VDC x .02a.** You will need a .18-watt resistor, so if you use a 1/4-watt resistor you'll be fine (1/4 watt = .25 watt = 250 milliwatts).

#2 USING OHM'S LAW

Another way you might apply Ohm's Law involves a 12-volt relay and a coil with 200 ohms resistance. You need to know how much current is required. A relay, incidentally, is merely an electrically-operated switch for opening and closing a circuit. Invert the equation for Ohm's Law to read: **Current (Amperes) = Pressure (Voltage) / Resistance.** Using the information provided, you can figure out that **Amperes (I) = 12VDC / 200 = .06.** Therefore, this circuit will require 60 ma.

enough, you might actually feel the plastic hose begin to expand under the pressure. When electrical resistance gets too high, the wires heat up. This is the principle behind the heating element in your toaster.

As you begin building circuits for your robots, you will need to apply these basic principles of electricity. Perhaps the most important law of electricity—one that you will use in many different ways—is Ohms Law. George Simon Ohm, a German scientist working in the early 1800s, discovered the direct relationship between volts and amperes. Specifically, he determined that the resistance of a circuit (R) is equal to the voltage applied to a circuit (E) divided by the resulting current flow in amperes (I). In other words, **Resistance(R) = Voltage or Pressure (E) / Amount of Current (I).**

Don't let the equation worry you. Ohm's Law is quite simple; you can prove it by inverting the equation to read **Pressure = Amount of Current** x **Resistance** and referring to the water hose example. By stepping on the hose, you create resistance and can control the water's pressure directly. Move your foot slightly and notice how the distance the water shoots changes. Put more resistance on the hose, and the water does not shoot as far. Similarly, if your foot resistance remains constant while you turn the tap to vary the flow of water, the pressure also changes.

At first glance, a schematic drawing of a circuit appears to be very confusing. Learning Greek must be simple in comparison!

SCHEMATIC DIAGRAMS—ELECTRONICS ROADMAPS

At first glance, a schematic drawing of a circuit appears to be very confusing. Learning Greek must be simple in comparison!

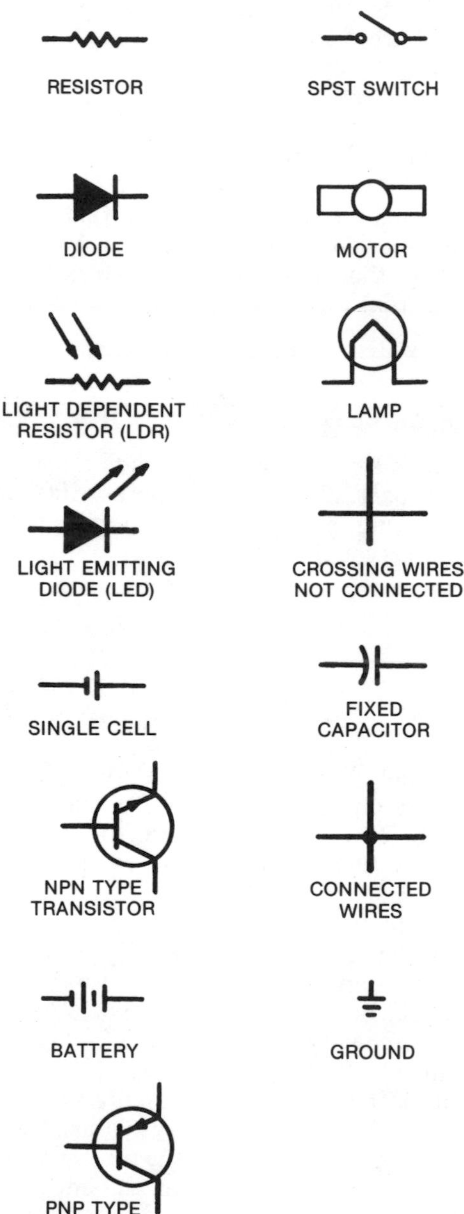

Symbols for electrical components used in schematic diagrams.

Actually, once you learn the various symbols used in electronic notation and feel comfortable with the principles of electricity and electronics, you are well on your way to reading circuits. A schematic drawing is your road map to a circuit.

Reading a circuit is just like reading a book. You read from left to right. Think of the left side of a circuit or component as the *input* (where the electricity flows in) and the right side as the *output* (where the electricity flows out). Remember, too, that electricity enters a circuit from the power source and must come full circle, returning to the original power source.

SOLDERING TIPS

Solder is an alloy of lead and tin that is used to connect wires and electronic components. It melts quickly and is easy to work with. There are several kinds of solder, but for electronics work use *only* rosin-core solder with a 63/34% mixture of tin and lead. This is a hollow wire with rosin inside. The rosin acts as a flux to keep the solder joint from oxidizing and insulating your circuits. If you have trouble with oxidation forming on your solder joints, increase the amount of flux by using some soldering paste.

Use a small, electric soldering iron. A pencil-type iron between 15 and 25 watts is best. The fine tip is just right for delicate work. New irons must be "tinned" before using. Plug the iron in; when tip is hot, melt a bit of solder on the tip. Run melted solder all over the tip, then carefully wipe it clean on a damp cloth. A thin coating of solder will remain on the tip.

When soldering, you should:

1. Make sure the surfaces you are soldering together are clean. If necessary, scrape wires with a knife blade or sandpaper or use a pencil eraser to remove oxidation.

2. Make sure you have established good contacts between the parts you are soldering. Do not use solder to fill in gaps between wires.

3. Apply your iron's hot tip to *all* surfaces being joined. Hold for a moment, then touch solder against the hot joint. Solder will flow into the joint. Remove solder, but hold iron on joint a moment longer, giving melted solder a chance to flow around the joint. Remove the iron and allow joint to cool. Keep joint still until your solder has cooled.

4. Use as little solder as necessary. Keep the joints neat and clean.

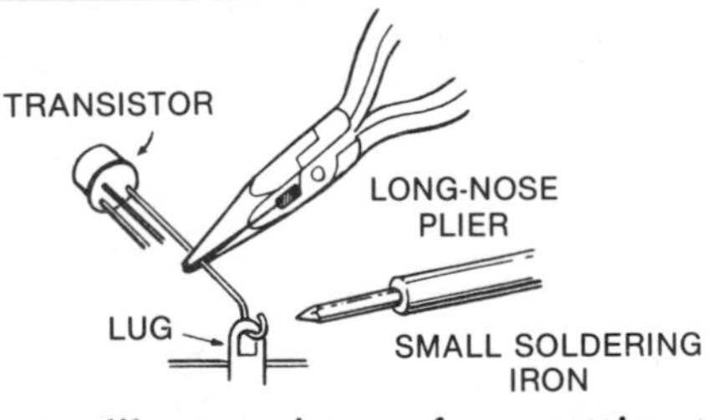

5. Keep sensitive components, like transistors, from getting too hot. If you clamp a heat sink or hold the wire lead between the component and the joint being soldered with a pair of needle-nosed pliers, the metal of the heat sink or pliers will conduct most of the heat away from the component. Your grandmother uses the same principle when she puts a silver spoon in her porcelain cup before pouring the scalding-hot tea. This keeps the heat from cracking the cool cup. Copper alligator clips (available at Radio Shack) make excellent, inexpensive heat sinks!

6. When finished soldering, melt a large glob of solder onto the tip of your iron and let cool. This will protect the delicate tip from accidental damage and prevent oxidation from forming on the tip of your soldering iron.

CAN YOU READ THESE ELECTRONIC ROADMAPS?

Below are two schematic diagrams. Can you describe what is happening?

1) Can you trace the flow of current through this circuit? This is the schematic for a common household item. What is it?

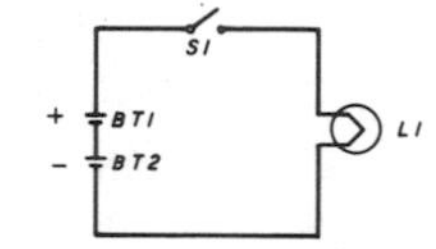

2) Trace the flow of current through this circuit with the switch (2) in position 1. What happens when the switch is moved to position 3?

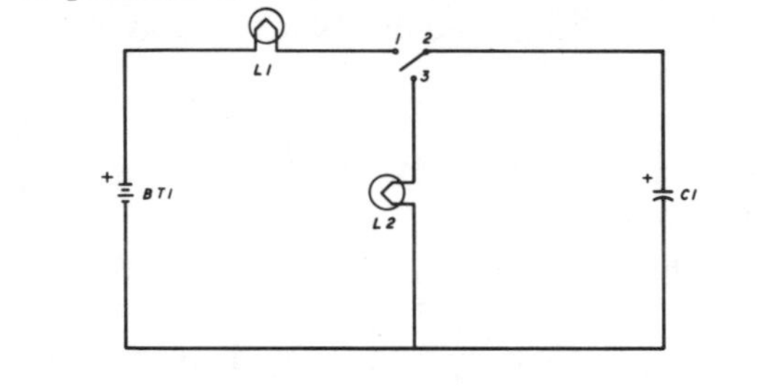

ANSWERS TO THE ELECTRONIC ROADMAPS QUIZ

1) When the switch is on and the circuit complete, electricity flows out of the batteries, through the light, and back around to the batteries. This is the schematic of a flashlight.

2) This circuit is a bit more tricky. It also demonstrates how a capacitor works. With the switch in position 1, electricity flows out of the battery, through the capacitor, to the light (L1) and back to the battery. In the meantime, the capacitor is charging, and when it is fully charged, it blocks flow of DC current. L1 goes off. Move the switch to position 3. The capacitor serves as a battery and will light up L2 briefly. When the capacitor is discharged, L2 goes off.

USING A MULTIMETER

As you design circuits for your robots, you will need to determine resistance in a circuit or component, the amount of current needed, and the wattage in a resistor. These are just a few of the ways you will use voltage and amperage. To take these measurements, you need a meter. A multimeter measures volts, ohms, and amperes (or milliamperes). It is often referred to as a VOM (volt, ohm, and milliampere) meter.

The older, analog meters convert electrical energy into a mechanical motion, moving a needle across a scale. A digital multimeter (DMM), on the other hand, converts the analog measurement into whole numbers and gives you a numerical readout on a liquid crystal display, much like a digital clock. The number or the point on the scale where the needle points indicates the amount of current (amperes) flowing through the meter and, therefore, your circuit.

As a voltmeter, the VOM reads both ac and DC voltage. Be careful to set your meter on the correct scale. *If you use the DC scale to read ac, you could ruin the meter.* The black probe (negative) is connected to the negative, or common, terminal. The red probe (positive) is connected to the positive terminal. *Reversing polarity and touching the positive probe to the negative lead and vice versa could also damage your meter.*

When using the VOM to read amperage, follow the same precautions as with voltage readings: Observe polarity and use the correct scale. Also, do not measure a current value that is higher than the VOM's range. If you are in doubt about the amperage, set your meter on the highest range to start. You can always switch to a lower range to get the most accurate reading. If necessary, you can also protect your meter by putting a .1 ohm resistor between the circuit and your meter. To account for the resistor, multiply your meter readings by 10.

To check resistance, set the multimeter to R x 1. Then, touch the two probe tips together. Notice that the needle (or readout on a DMM) will register close to zero. Still holding the probes together, slowly move the ohms adjustment dial until the needle or readout registers exactly zero. With the meter registering zero resistance through itself, you can now measure the resistance in a circuit or individual component. Experiment with the meter and measure the ohms in several resistors by touching the probes to the components' wire leads.

RESISTORS

Resistors are placed in an electronic circuit to provide resistance in the circuit in order to 1) reduce or control the flow of current, 2) cause differences of voltage at certain points, or 3) reduce the voltage applied to a device. The more resistance in a circuit, the less current it will let pass. The degree of resistance will vary depending on a component's composition. The unit of resistance is an ohm.

Few resistors are marked with numerical values. Rather, they are color coded, with bands of color encircling the resistor. Each color represents a number. To determine the amount of resistance, hold the resistor with the bands to your left. Reading from left to right, the first two bands are the first two numbers of the resistance. The third band tells the multiplier or number of zeros to add. The fourth band indicates the accuracy of a resistor.

Referring to the chart below, this is a 100 ohms resistor:

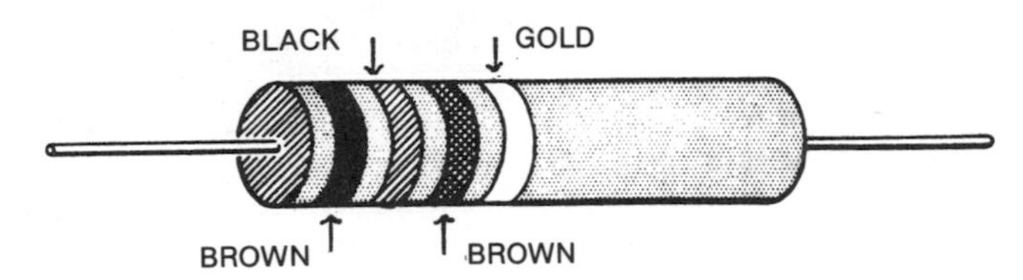

The brown band is one, the black band zero. The third band is also brown, so multiply by 10. 10 x 10 = 100. The fourth band is gold, indicating that the resistor may vary by 5% plus or minus. Had it been silver, the resistor could vary by 10% and without a band by 20%.

FIRST NUMBER		SECOND NUMBER		MULTIPLIER	
BLACK	0	BLACK	0	BLACK	1
BROWN	1	BROWN	1	BROWN	10
RED	2	RED	2	RED	100
ORANGE	3	ORANGE	3	ORANGE	1,000
YELLOW	4	YELLOW	4	YELLOW	10,000
GREEN	5	GREEN	5	GREEN	100,000
BLUE	6	BLUE	6	BLUE	1,000,000
VIOLET	7	VIOLET	7		
GRAY	8	GRAY	8		
WHITE	9	WHITE	9		

You can determine the capacity (wattage) of a resistor by its size. The resistors illustrated below are drawn to size. For most of your electronics work, 1/4- and 1/2-watt resistors will be adequate.

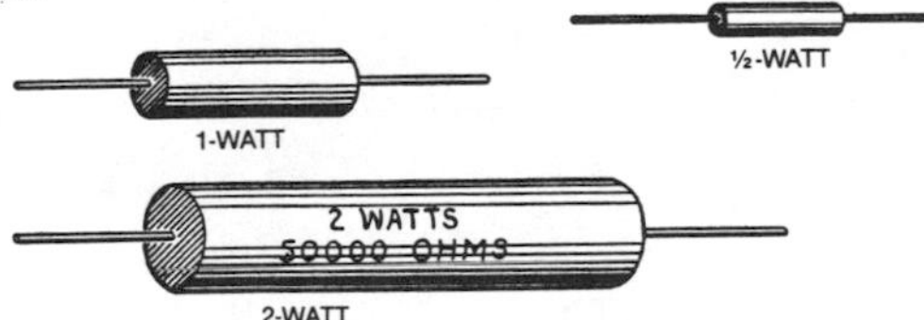

Being able to determine resistance and wattage of a resistor just by looking at it is an impressive skill. One way to help you acquire that skill is to memorize the following phrase: **B**ig **B**oys **R**arely **O**pen **Y**esterday's **G**rapejuice **B**ut **V**isit **G**reat **W**ineries. Note that the first letter of each word corresponds with the first letter of each color in resistor code.

TOOLS RECOMMENDED FOR ROBOT PROJECTS

15 - 25 watt pencil-style soldering iron
The low wattage of this iron will help to protect delicate components from burning out. You can buy a good iron for around $10 to $15. **(1)**

Rosin-core solder
This solder is 63% tin and 34% lead. The rosin core serves as a flux to prevent the wire or metal leads on components from oxidizing and insulating the joint. The rosin forms a coating that protects the hot metal from air. Do not substitute acid-core solder. It will corrode the metal. A spool costs $1-$3 and lasts a long time. **(2)**

Wire strippers
This clever tool strips the insulation off of wire without cutting the wire itself. Good wire strippers are designed to accommodate a wide range of wire sizes. Prices go from $2.50 to $9. You can also strip off insulation with a good pair of wire cutters or a pocket knife, but take care that you do not cut into the wire. **(3)**

Diagonal wire cutters
Diagonal cutters (commonly referred to as "dikes") are used to cut wire, trim leads, and strip insulation. They cost between $3 and $10. Good cutters are important, but if you need to save money, select a pair of needle-nose pliers that have cutters included. **(4)**

Screwdrivers (several sizes)
In addition to having both a standard and phillips screwdriver, you will want a set of miniature screwdrivers, such as the ones watchmakers use. A set costs around $3. **(5)**

Needle-nose pliers
Also called long-nose pliers, these are used to hold wire and small components in place, bend and splice wire, and to divert heat away from delicate components. You can buy needle-nose pliers for around $6. **(6)**

Black vinyl electrician's tape
This plastic tape is a good insulator. You can wrap it around exposed wire when you are concerned that the joint or wire end might come in contact with another metal

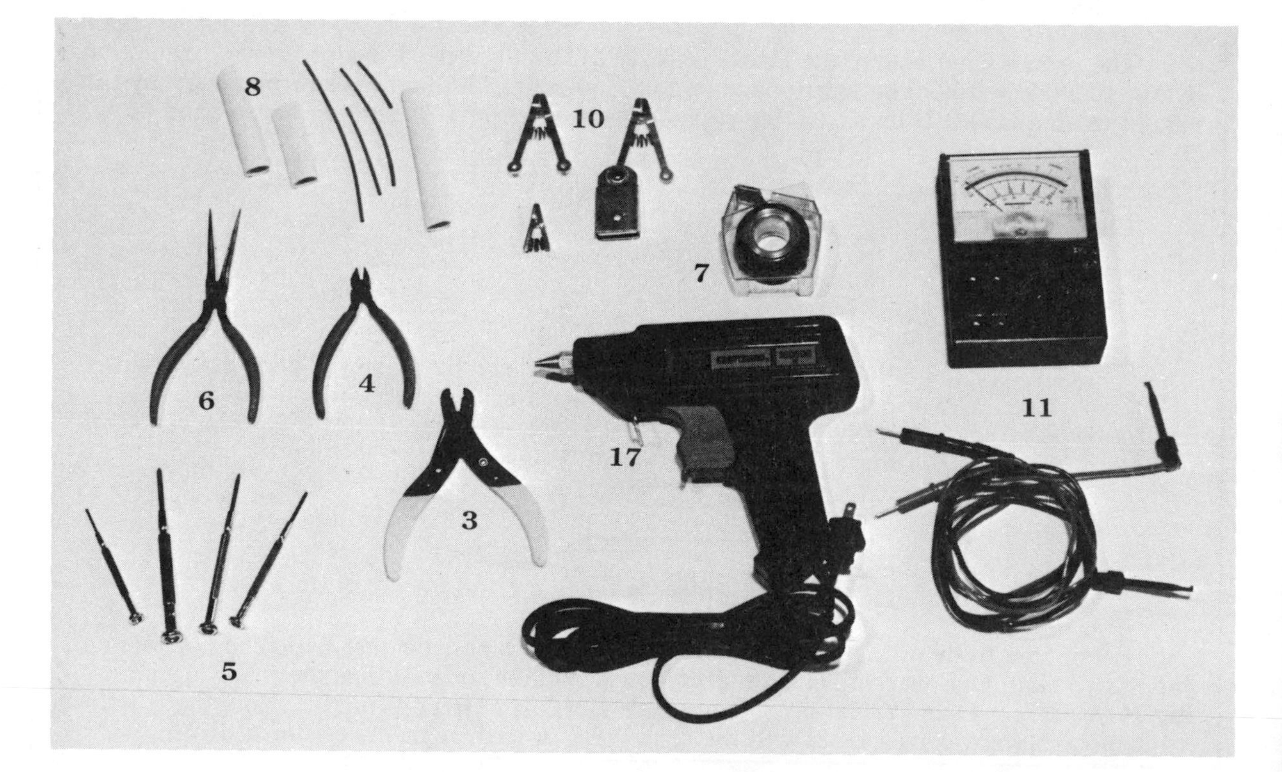

surface and short your circuit. It is more commonly used with large-scale electrical wiring, not electronics. A roll costs about $.80. **(7)**

Heat shrink

This special plastic tubing is slipped over an exposed wire junction as insulation. Once in place, apply a little heat (match or soldering iron) and the tubing shrinks to fit snuggly around a wire. It is available in assorted sizes and colors at any electronics store and costs $1 to $2 per package. **(8)**

Large soldering iron

Also called a soldering gun, this variety has a wattage that ranges between 100 and 300 watts, sometimes more. It is too hot for soldering delicate components, but useful in soldering pieces of metal together when constructing your robot's body. This is not one of the first tools you need to buy, but eventually you will probably want one. They range in cost from $20 to $75. **(9)**

Heat sinks (optional)

These metal clips resemble a small pair of pliers. They are made of highly conductive metal and used in soldering. When clipped on to a wire or metal lead on a component, they divert heat away from the component. A set of variable sizes costs about $3. You could substitute some inexpensive, copper alligator clips. **(10)**

Soldering paste

This noncorrosive flux is another way to prevent oxidation from insulating a circuit joint. Furthermore, if you rub paste on copper wire, you can make your own inexpensive soldering wick that helps collect solder when you are trying to desolder a joint. A tube costs $1 to $2.

Extra-large toenail clippers

If you can find a pair of big toenail clippers—about 4'' long—buy them. They are invaluable tools for all sorts of jobs. The best ones have a cutting edge with a convex curve. They cost around $1.50.

Multimeter (VOM)

This electronic measuring device lets you probe points in a circuit to determine

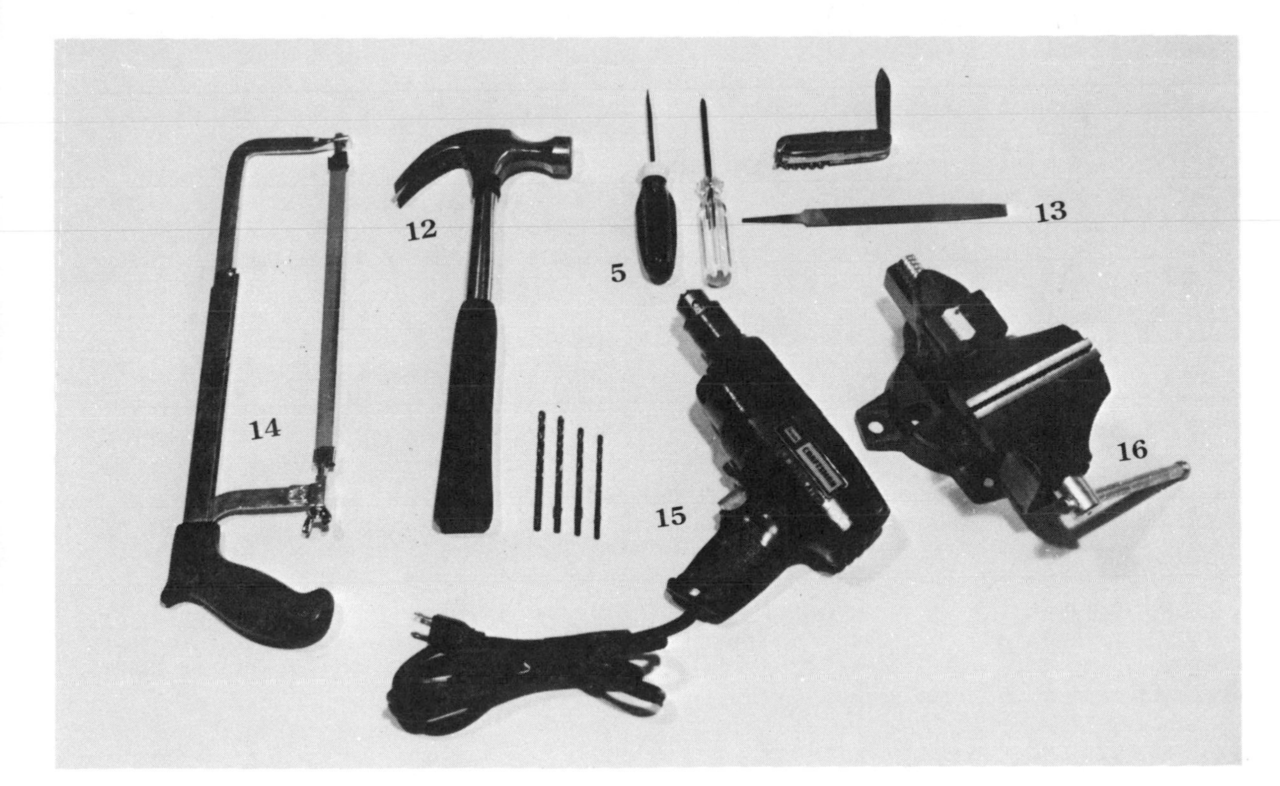

resistance (ohms), voltage, and amperage. You can get an analog meter (needle-type) for around $10 to $15 or a newer, digital meter for between $50 and $75. Sometimes you can pick them up at flea markets. Also, watch the ads in *Computers and Electronics* magazine. **(11)**

Hammer

This needs no explanation. It is especially handy for driving axles into wheels and working on wooden platforms. **(12)**

Files

Preferably, you should have several with different degrees of coarseness. If you only get one, select a medium coarseness. **(13)**

Hacksaw

This special saw is designed to cut metal. It will cut off the tops of bolts in no time. Use this saw to modify the base of the furniture caster (front wheel) in the Moth. **(14)**

Electric drill

Whether you make your robot's platform out of wood or plexiglas, you will need to drill some holes in the base. While an old rotary manual drill will work, you'll be better off with a hand-held electric drill. If you don't have one around the house, you can pick up one at a flea market. Or, watch the classified ads in the newspaper. Used drills should cost about $15; new ones about $30. **(15)**

Vice

When you are sawing wood or metal or drilling holes, you will find a vice helpful. It holds objects perfectly still while you work. If you have any kind of a workshop, you probably have a vice. A heavy, metal vice costs around $15, but you could find a bargain at a flea market. **(16)**

Hot glue gun (optional)

This tool looks and works like a pistol-style soldering iron, except that instead of solder you use solid sticks of glue. You can glue small parts onto your moth without screws and bolts. A glue gun costs about $10. **(17)**

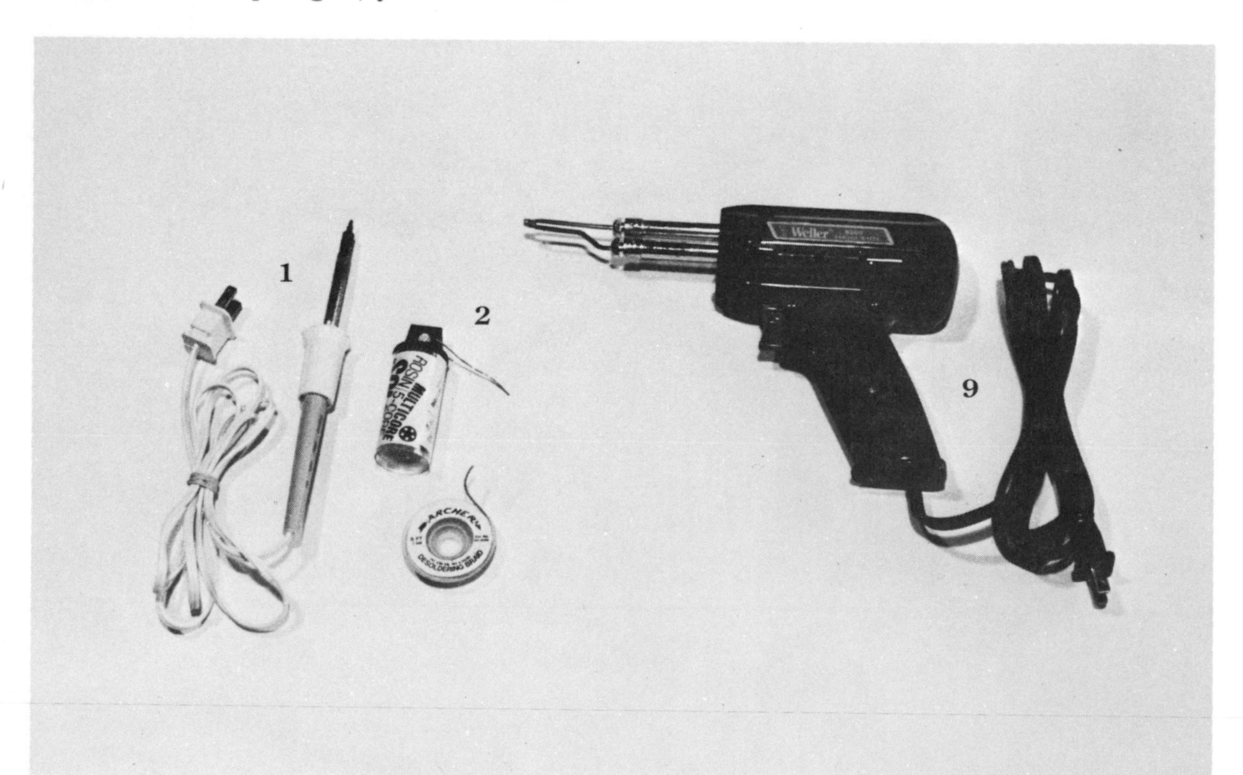

CAPACITORS

This electronic component is used to block the flow of direct current (DC) and actually store electricity. All capacitors, no matter what the size, are basically the same. They have two conducting metal plates (usually aluminum foil) separated by a layer of insulating material (air, oil, paper, mica). While the distance between the metal plates is microscopic, the insulation is so effective, it might as well be the Grand Canyon.

When connected to a DC power source, charged particles flow from the plate connected to the positive terminal, through the battery to the plate connected to the negative terminal, where they collect. The capacitor is fully charged when voltage across capacitor plates equals voltage across battery terminals. It holds a charge like a battery; unlike a battery, the charge can be released instantly. Short out the two plates with a conductor and the electrons on the negative plate are discharged with a surge! *Be careful that you do not touch a charged capacitor and get an electrical shock.*

Capacitance refers to the amount of charge (number of charged particles) required to build up a certain potential difference. (Remember that potential refers to voltage or the potential for pressure—a condition that causes current to flow.) Think of a capacitor collecting electricity the way a dam collects water. When the dam opens, the water rushes out with great force. Similarly, a capacitor can discharge with a surge of energy. You can increase capacitance by increasing the size of the plates (like making the walls of the dam higher) or decreasing the distance between plates. The unit of measurement for capacitance is called FARAD. A farad is the number of COULOMBS (quantity of electricity in a charge; similar to using the term gallon to measure a quantity of water) required to charge a capacitor to a voltage or potential of one volt. **Farads = Coulombs/Volts.** A farad is too large a unit of measurement in electronics; we use a smaller unit (microfarads—μF), which is equal to one millionth of a farad. You might note that a **Farad X Ohm = Second (C x R = Time).** This principle is often used in making oscillators or setting the flash rate in blinking lights.

N-P-N vs. P-N-P TRANSISTORS

Here's an easy way to remember which way the current flows through a transistor and, therefore, which way the emitter's arrow should point on the schematic drawings.

On the P-N-P transistor, the arrow points in. P for P-N-P; P for "pointing in."

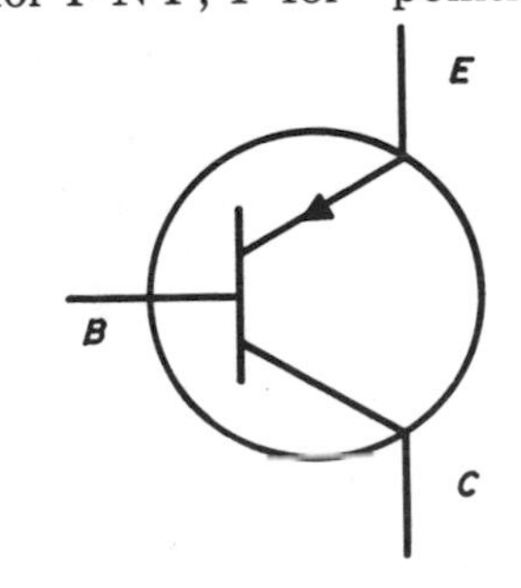

On the N-P-N transistor, the arrow points out. N for N-P-N; N for "not pointing in."

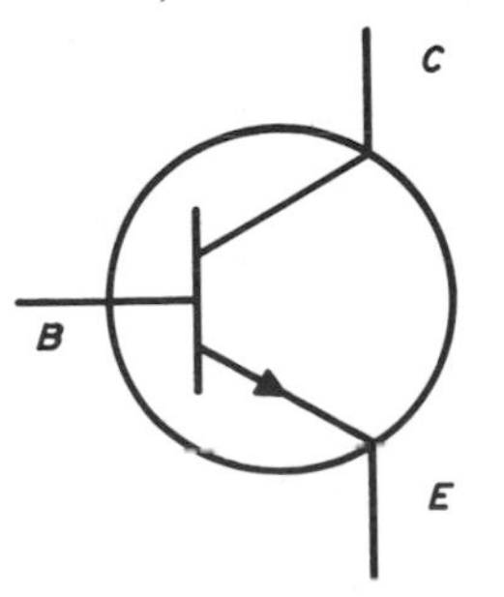

RELATION EQUATIONS

With a little math, you can figure out a lot about your robot. Here are some of the equations you can use:

Ohms Law	$E = I \times R$	Volts = Amps x Ohms
Energy	$J = E \times Q$	Joules = Volts x Coulombs
Energy	$J = W \times T$	Joules = Volts x Seconds
Power	$W = E \times I$	Watts = Volts x Amps
Charge	$Q = E \times C$	Coulombs = Volts x Farads
Charge	$Q = I \times T$	Coulombs = Amps x Seconds
Time	$T = R \times C$	Seconds = Ohms x Farads

Energy (described as the ability to do work) can be measured in many ways. For example, 1 foot - 1 pound = 1.358 joules. This means that a 1 pound object dropping about a foot releases a little over 1 joule of energy. Or, it takes about 1-1/2 joules of energy for your robot to lift 1 pound about 1 foot.

Power is the rate of energy use in watts. Since a joule per second equals a watt, your robot needs 1.5 watts of power to lift 1 pound about 1 foot in 1 second. Using a 1.5-volt battery, the equation Watts = Volts x Amperes tells you that you will need at least 1 ampere of current to get enough wattage.

Once you have the wattage, you can also figure out the horsepower (HP=746 watts). By dividing 1.5 by 746, you know that the hypothetical 1.5 watts robot arm has .002 HP (or 2/100 ths of a horsepower).

Charge is measured in coulombs (Q), and 6.24×10^{18} = 6,240,000,000,000,000,000 electrons make up 1 coulomb of negative charge. It is hard to measure directly, but you can figure out the charge by knowing the voltage on a capacitor. The equation is $Q = E \times C$, where C refers to the size of the capacitor in farads. ($Q = I \times T$) will also give you the charge. This equation can be used to measure the amount of metal deposited on an object during an electro-plating process.

Finally, should you be working with flashing lights or sound frequency oscillators, you are going to be interested in time factors. How often does a light blink? What are the attack and decay rates of a sound? One equation you can use is $T = R \times C$.

You may notice throughout this book that sometimes voltage is labled E ($Q = E \times C$) and other times V ($R = V / I$). In electronics, use E when you are referring to voltage in a circuit. Use V when you are referring only to the voltage across a single component.

TRANSISTORS

Transistors are triode (3 electrode) semiconductors, first developed at Bell Telephone Laboratory in 1948. They have virtually replaced the earlier vacuum tubes because they are smaller, less expensive, longer-lasting, and much more efficient. Further, they consume practically no current, so batteries last longer. Transistors are used to control current. In circuits, they can serve as amplifiers, detectors, relays, modulators, and more. They can even be used to change ac into DC electricity!

Whereas a diode is a one-way conductor, a transistor consists of two diodes, back to back. The most common is the P-N-P type junction transistor. It has three, thin layers of silicon or germanium (a special, somewhat scarce metal). These layers are called the EMITTER, BASE, and COLLECTOR. The emitter and collector are positive (P)-type silicon, while the base is negative (N)-type and is sandwiched between the other two. In an N-P-N transistor, the charges are simply reversed.

A wire lead is connected to each layer. In both an N-P-N and P-N-P transistor, current enters through the emitter lead and leaves through the collector. For your purposes, the most important distinction between the two types of transistors is the nature of the electrical charge on the layers. Therefore, you *must* get the leads hooked up to the battery correctly. When building the Moth, carefully read the directions for connecting the transistors. Sometimes transistors are marked so that you can recognize the collector lead. More often you will have to refer to the diagram and instructions included with the transistor. Don't lose this paper. Keep it with your extra transistors.

In using transistors:

1) Do not cut the leads too short.

2) Comply with indicated pin assignments, otherwise you will blow out the component.

3) Connect the transistors in your circuit after everything else is connected.

4) Avoid getting the transistor too hot while soldering; use a heat sink. (See the tool list for more information, page 39.)

BUILDING THE MOTH—A LIGHT-SEEKING ROBOT

We all have observed that moths are attracted to light. We've watched moths and other small insects flit about a kerosene flare. They are drawn instinctively to the light and begin a dance of death that ends when they fly straight into the fire.

As a robot builder, you might wonder, if you can duplicate something like the moth—something that responds to light. Let us consider a moth to be a motorized, mobile platform with light sensors that steer it toward maximum light. Batteries are needed to power the robot. A couple of small transistors will amplify the signals that the light sensors generate in order to create enough power to drive the moth's motors. The gearhead motors, in turn, drive the wheels.

Let's start with the complete parts list on the next page. We have used Radio Shack parts when possible. Note that our list is for a basic moth. As you become more familiar with robot building, you may wish to change the design or add special features, such as LEDs.

Illustration by Bob Johnson

SHOPPING LIST

1. Triangular platform (wood, plastic, any non-conductive surface) about 1/8" thick and 8"x8"x8"—The advantage of plexiglas is that most metal and plastic objects (except polyvinyl chloride) will adhere with a super glue (e.g., Crazy Glue). The plastic pegboard screen in the Lite Brite toys works well too. All the screw holes have been drilled. *WARNING—do not try burning or melting polyvinyl chloride plastics. They give off a poisonous gas.* Colored plexiglas costs about $7 (includes cutting).

2. 1 small, **protoboard** (also called a breadboard or experimenter's board)—Minimum size is 12 holes x 4 holes; you may want one slightly bigger to allow for future circuit experimentation. The Radio Shack board is a bit too large. If you cannot find a smaller one, you can cut it to size. BEFORE CUTTING: Strip part of the backing off and remove the metal clips in the columns you plan to cut through. A small board costs about $4.

3. Super glue (e.g., Crazy Glue)—If you are using a plastic or plexiglas platform, you can attach most components with glue; $1.50.

4. 2 plastic **battery holders** (for 4 C batteries)—Radio Shack #270-385, $.99 each.

5. Package each of #4-40 x 1/2" **flat head steel machine screws**; #4-40 x 1/4" x 3/32" steel **hex nuts**; #4 steel **split lock washers**—When assembling, put washer on just before nut. This helps keep screws from loosening.

6. Small, furniture caster for front wheel, about 1-1/8" diameter—If you cannot scrounge one, go to a hardware store. Make sure it has a flat base, not a shaft, so it can be screwed or glued to the Moth's platform.

7. 2 **gearhead motors** 3-6 VDC (volts DC) and gear units from 2 radio-controlled toys—These are the very best and easiest to use.

8. 2 **disc capacitors** (.05 μF)—Don't buy capacitors, if the motors (above) already have them. See step 7A for an explanation. Radio Shack #272-134, $.85 for two.

9. Brass strip .064 x 3/4"—Hobby store $1.25.

10. 3 steel **flat head screws** (6/32 x 2"); #6.32 steel hex nuts; #6 steel split lock washers—Hardware store $.50.

11. Brass tubing 3/16"—A 7-8" piece will do; hobby store $.45.

12. 2 **toy car or airplane wheels, 1-½" d**—Scavenge these from radio-controlled toy or buy at hobby store for around $2.50.

13. Plated brass Dura-collars 3/32"— Hobby store $.80 for four. If you can't find them: Du-Bro Products Inc., 480 Bonner Rd., Wauconda, IL 60084.

14. Double-sided adhesive foam strip—The foam should be very thin; hardware or hobby store $1.15.

15. 1 **index card or a thin piece of plastic, such the plastic lid on a box of greeting cards or stationary**—This will be used to make blinders for the photocells.

16. 2 **cadmium-sulfide photocells**— Radio Shack #276-116, $1.29 each.

17. 1 small bottle of **flat black enamel paint** (model-builder's paint)—Hobby store $.49.

18. 2 **general-purpose, amplifier transistors** Radio Shack #276-2014, $.89 each, or #276-2009, $.79 each.

19. 1 **1/4 watt, 100K resistor** (brown-black-yellow)—Radio Shack #271-1347, $.39 for 5.

20. Insulated, solid-core copper wire, 22-gauge—Radio Shack #278-1295, $2.19 for 100 ft.

21. Slide switch SPST (single pole-single throw) 3A @ 125 VAC—Radio Shack #275-401, $.89 for two.

22. 4 **nickel-cadmium (Nicad) batteries** (C size)—Any electronics store, $6.50 for 2. You can use regular batteries (C size); you'll just have to replace them more often.

23. Square plastic stick 1/8"—Hobby store $.40.

24. Small piece of **foam rubber** to serve as a bumper on the front of the Moth—Scrounge something around the house.

BUILDING THE MOTH

In describing the location of items on the moth, we will be looking down on the platform with the forward angle pointing away from your body.

WIRING IN SERIES AND PARALLEL

When working with electricity, you must always consider the polarity of the components in your circuit. Do you want to hook up the negative lead to the positive or vice versa? In our Moth project, most of the wiring is in series.

In a *series* circuit, you connect the positive to negative, positive to negative, positive to negative, and so on. The same current is flowing through one loop of a circuit. Think of a string of Christmas tree lights. If they are wired in series and one light burns out, they all go out. The circuit is no longer complete.

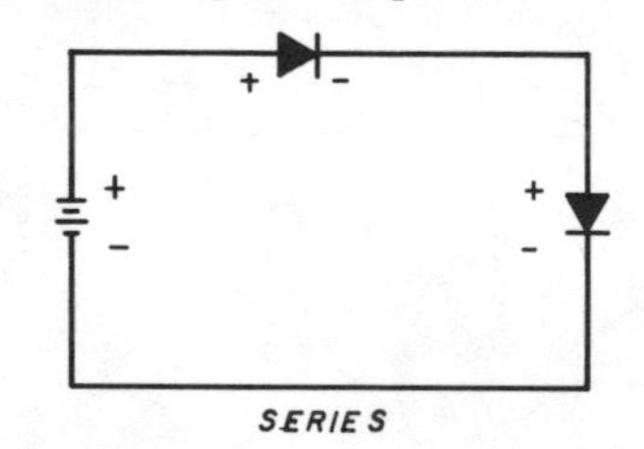

SERIES

In a *parallel* circuit, you connect the positive to positive, positive to positive, positive to positive, and negative to negative, negative to negative, negative to negative, and so on. The voltage will read the same throughout one loop of the circuit. If the tree lights are wired in parallel and one light burns out, the rest remain lit. Current can still flow through the circuit.

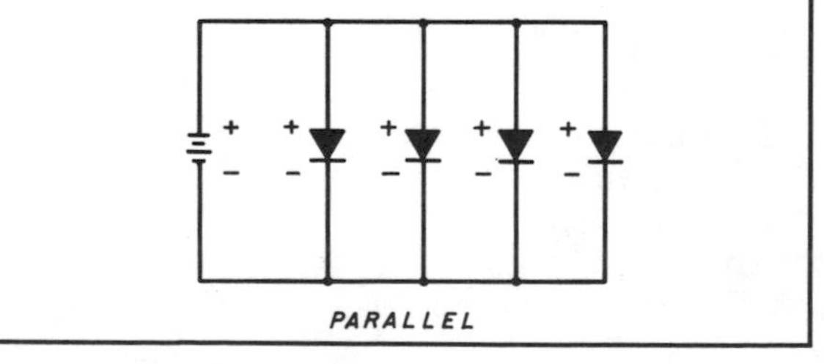

PARALLEL

1. THINK AHEAD

A) An 8" equilateral triangle should be large enough to hold all of the components. But just in case, cut out an 8" triangle from a piece of paper or cardboard and practice laying out all the parts.

B) Read all of the directions before starting to build. You'll have a better chance of

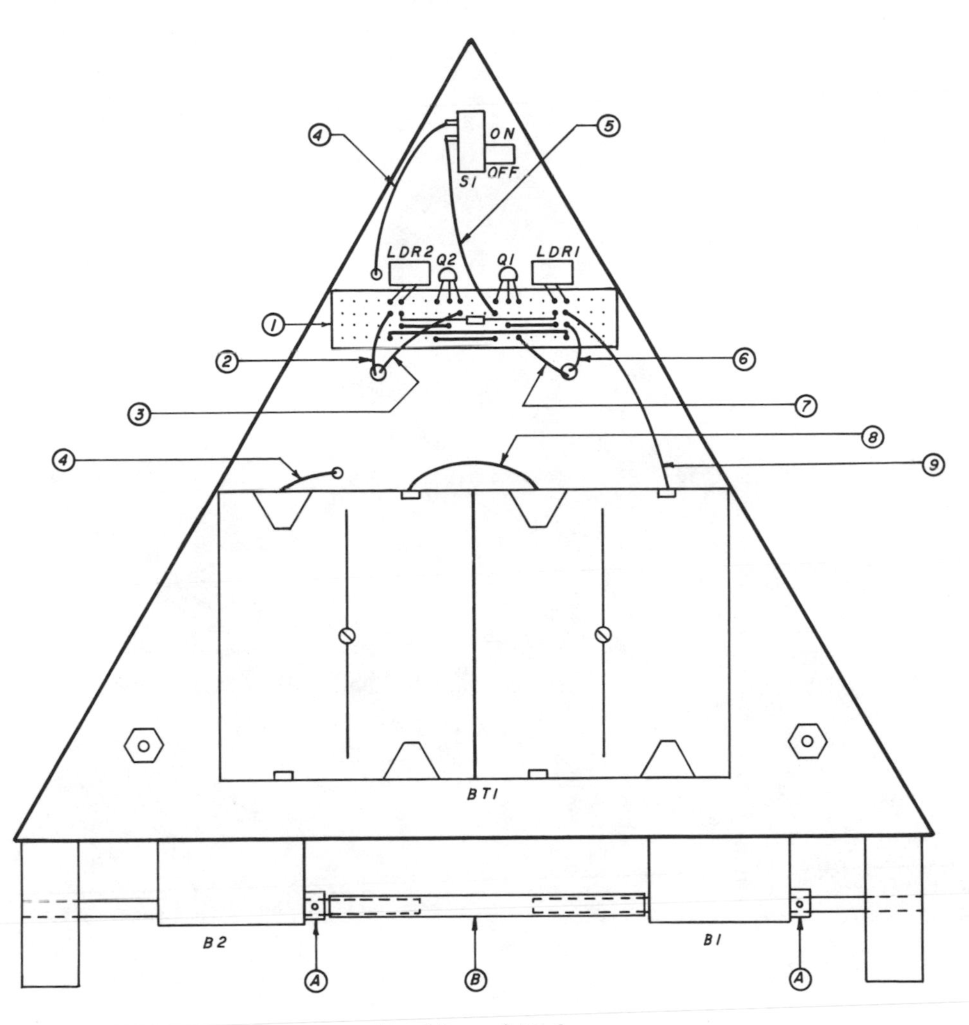

Top View of Moth

NOTES:

1. BREADBOARD CIRCUIT (SEE DETAILED DRAWINGS)
2. LEAD FROM B1 TO LDR2
3. LEAD FROM B1 TO COLLECTOR Q2
4. NEGITIVE LEAD FROM BT1 TO S1
5. LEAD FROM S1 TO EMITTER
6. LEAD FROM B2 TO LDR1
7. LEAD FROM B2 TO COLLECTOR Q1
8. CONNECT INSIDE POSITIVE AND NEGITIVE LEADS
9. POSITIVE LEAD FROM BT1 TO LDR1

(A) COLLAR
(B) SLEEVE

catching those steps that may snag you up later.

2. CUT THE PLATFORM

If all the parts appear to fit on an 8'' triangle, go ahead and have the plastic or wood platform (#1) cut.

3. MOUNT PROTOBOARD

A) On the top of the platform, mount protoboard (#2). Mount board about 2'' from front of platform. If using plexiglas or plastic, adhere the protoboard to the platform using a super glue (#3).

B) *If your platform is wooden,* place pro-

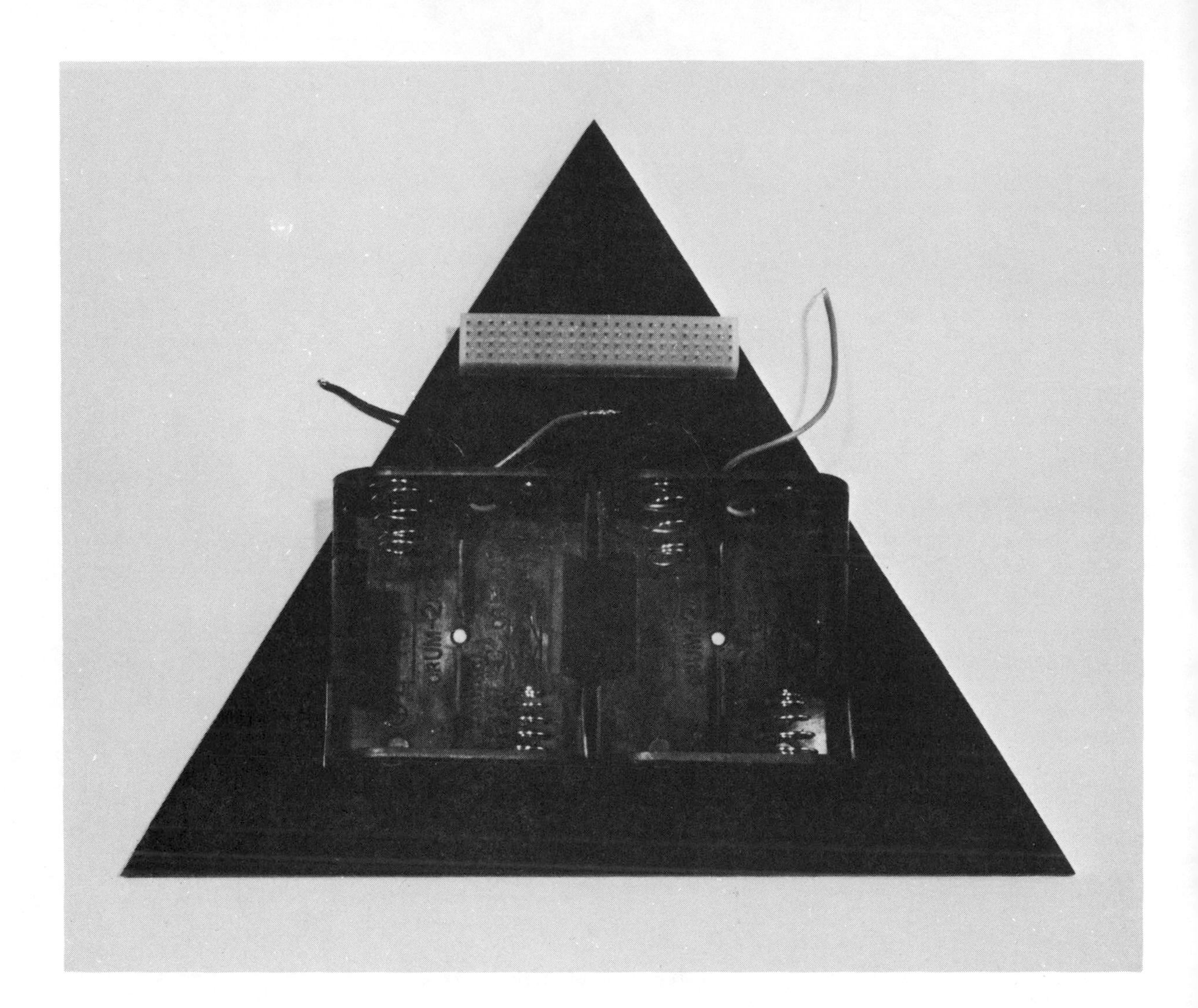

toboard on platform, orient screw holes so they go through the outermost column at each end of board. Mark underside of platform accordingly and remove board. Use a 3/32'' bit to drill holes in platform. REMOVE THE METAL CLIPS FROM THE OUTERMOST COLUMN ON EACH END OF PROTOBOARD. (THE SCREWS GO THROUGH THESE COLUMNS). PEEL BACK THE PROTOBOARD'S BACKING TO WHERE YOU CAN PRY CLIPS OUT USING A SMALL SCREWDRIVER. TRIM BACKING SO THAT THESE OUTSIDE COLUMNS ARE EXPOSED. Place protoboard back on platform and run 2 small (about 1/4'' self-tapping or wood screws up from underneath the platform. Screw down until snug.

C) Another technique on wood—Catch the board snuggly between a couple of round-head wood screws. Just be sure that the lip of the screw head catches the edge of the protoboard.

4. MOUNT BATTERY HOLDERS

A) Mount plastic battery holders (#4) with red and black leads facing forward. You will have to bolt them to the platform. Orient them so that their edge is flush with the back edge of the platform. Leave about a 1/4'' gap between the two holders. Mark platform so that you can drill a hole in the platform that

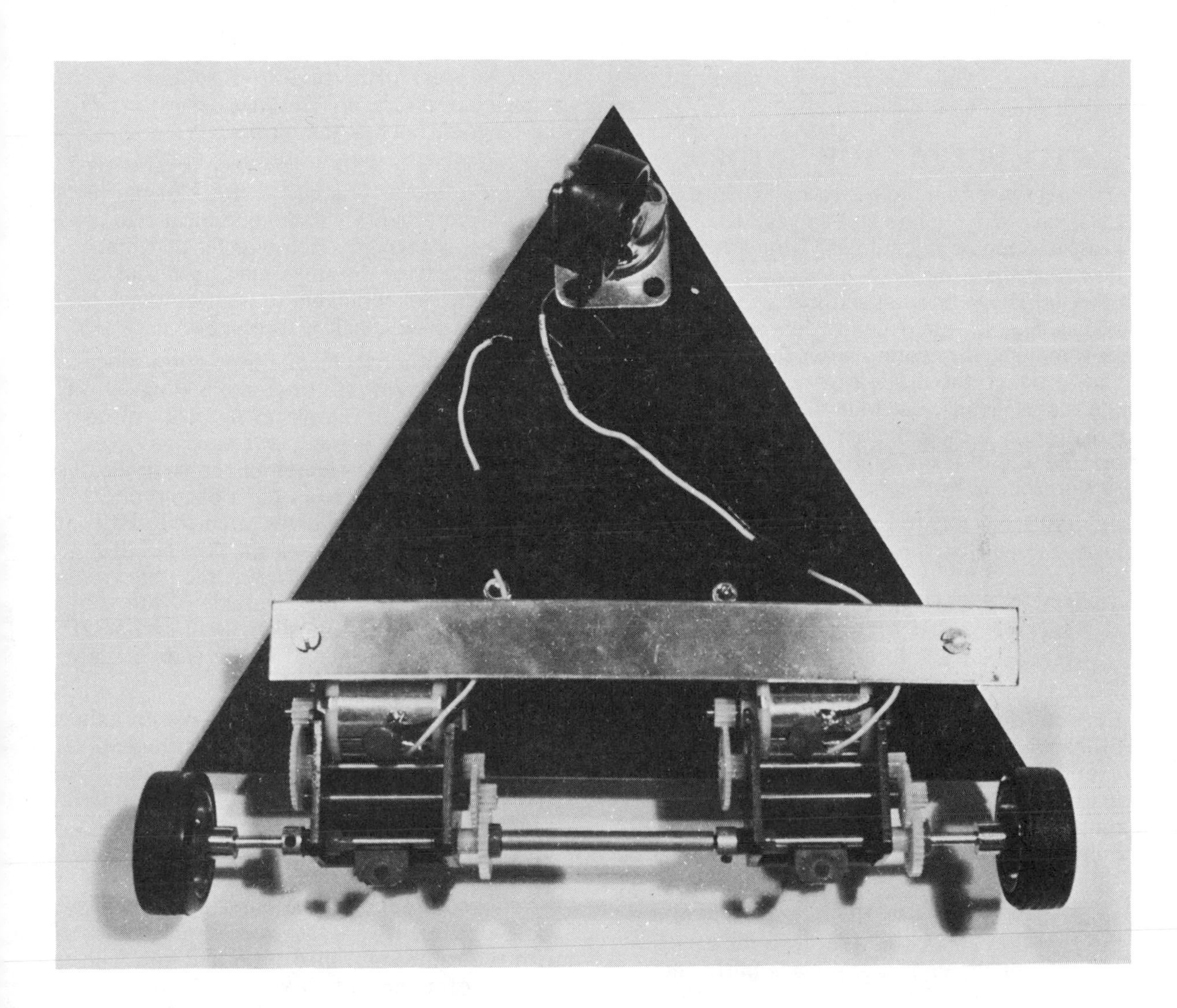

goes through the center hole in each battery holder.

B) Remove battery holders and drill holes using a 7/64'' bit. Place holders back on platform and run machine screws (#5) down from the top. Secure with washers and nuts (#5). Wire the battery holders in series by connecting and soldering only the two *inside* leads (one red and one black). DO NOT CONNECT THE OUTSIDE LEADS AT THIS TIME.

5. MOUNT THE FRONT CASTER WHEEL

A) Refer to Drawing 2 for orientation. If the caster (#6) has a square or rectangular mounting base, you may want to use a hacksaw to taper its base to the shape of the Moth's platform. This way the caster base will not protrude out from the platform. Or, just mount it a little farther back so that the entire base is hidden under the platform. If using plexiglas or plastic, glue the caster to platform. Be careful not to get glue in ball bearings.

B) If you need to use screws, you will have to select a machine screw and nut of the appropriate size. (Since we cannot know what brand of caster you will buy, it's impossible to tell you what size screw and drill bit to use.) Follow directions for orienting and drilling holes as listed above.

6. SALVAGE THE TOY MOTORS

A) Once you have two radio-controlled toys, take them apart and salvage the gearhead motor assemblies. *Save other parts from the cars—you never know when they will come in handy!* If a capacitor is connected across the two motor leads, leave it in place. In removing the motors from the toys, if you can't salvage the original wires running from the motors leads, cut four 6" pieces of wire (two for each motor), strip 1/4" of insulation off one end of each. Solder the stripped ends to the four motor leads.

B) Hold a battery across the two motor leads to see that the motor runs.

7. MOUNT THE TWO GEARHEAD MOTOR ASSEMBLIES TO THE MOTH'S PLATFORM.

A) Gearhead motor assemblies will vary from toy to toy. Some will have small brackets, others will completely enclose gears. Therefore, we can only offer general instructions for mounting the gearhead motors. You may have to be inventive, as all homebrewers must sooner or later.

B) If your motors do not already have a small capacitor across the motor leads, use the two disc capacitors (#8) (one for each motor), cut the capacitors' leads fairly short and solder across the motor leads, right where they attach to the motor. This routes any noise or interference (produced by the motors' brushes) around the motors. Ac current passes through the capacitors, circumventing the motors. A small amount of DC current is used to charge the capacitors. Once charged, all DC current flows through the motors. *Ceramic disc capacitors do not have polarity and can be mounted in either direction.*

C) Mount the gearhead motors underneath the Moth's platform. Exact placement will depend on the motors, but generally you will locate them toward the rear of the platform. The axle can even extend about 3/4" past the trailing edge of the platform. The units also must be spaced widely enough apart to prevent the tires from rubbing against the platform. You may want to trim the back corners of the platform.

D) Once oriented, you may be able to attach the motor brackets directly to the platform. If you use this technique, drill two 7/64" holes through each bracket and platform. Otherwise the units may move out of alignment. Bolt brackets to platform with machine screws, lock washers, and nuts (#5).

E) Another technique is to cut a piece of the brass strip (#9) to create a sling holding the motors in place. A 6-1/2" piece ought to be long enough, *but measure before cutting.* Make sure enough of the strip extends beyond the motors for a couple of 2" screws (#10) to come down from the platform through the ends of the strip, thus holding the whole unit in place. Orient the motors, run the strip along the underside of the brackets— creating the sling. Drill two 5/32" holes in the platform and in the ends of the strip. DO NOT ASSEMBLE.

F) Most gearhead motors have an axle that extends out both sides of the bracket. This makes it easy to use the units interchangably. Cut a 2" piece of brass tubing (#11) and slip it over the two inside axles (length may vary). This sleeve adds greater stability to the axle unit. Slip the wheel/tire assemblies (#12) onto the outer axles—you want a tight fit. Use the wheels that came with the radio-controlled toys, and you will be assured of a snug fit. If you need wheels with a larger diameter, try airplane wheels.

G) If the axle slides from side to side and disengages from the gears, slip a couple of collars (#13) on the axle. Exact placement depends on the design of your gearhead motors. The collar has a small screw in its side. Slide the collar against the motor bracket and tighten.

H) ASSEMBLE THE UNIT. If your motor leads touch the brass strip, they will short out. Run a length of black electrical tape along the strip. Cut two 3/4" pieces of adhesive foam (#14) and attach to brackets or motors as a cushion between platform and motors. Reorient the motors. IF THE SLEEVE (#11) BETWEEN THE TWO IN-

SIDE AXLES DOES NOT MOVE FREELY, THE MOTORS ARE NOT WELL ALIGNED. With motors in place, put the strip across. Run screws (#10) down from the top. Secure with washers and nuts. If the screws are too long, mark length, remove one at a time, and cut with a hacksaw. Keep the nut on the screw as you cut. Remove nut after cutting to clean screw's threads.

I) If the brass strip bows out too much in the middle, you can drill another 5/32" hole in the middle of the strip and a companion hole in the platform. Catch the middle of the strip with another machine screw (#10).

8. PREPARE BLINDERS FOR PHOTOCELLS

A) Using an index card or thin piece of flexible plastic (#15), make blinders to go around the two photocell resistors (#16). These will help block peripheral light sources from interfering with the Moth's operation. The pattern below is sized for Radio Shack photocells.

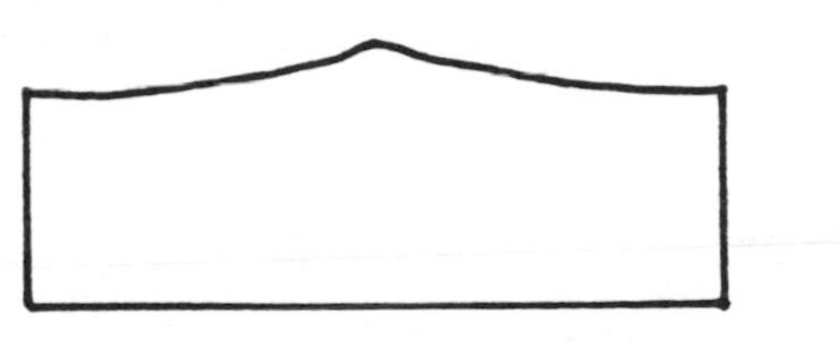

Pattern for Blinder (full size)

B) Wrap a blinder around each cell. The flat edge is flush with the back of the cell; the blinder is widest at the top of the cell and tapers toward bottom. Fasten with tape. Trim tape where it extends beyond blinder. Once in place, paint the blinder (inside and out) and the back of the photocell with flat black enamel (#17). DO NOT PAINT THE PHOTOCELL'S WINDOW. Let dry.

9. WIRE THE PROTOBOARD

With the undercarriage in place, begin wiring the Moth's protoboard. Refer to the schematic, Drawing 3.

A) Insert two photocells (#16) into the protoboard. These are referred to as LDR1 and LDR2 on the schematic. The wire leads should go into adjacent holes at each end of the board's first row. MAKE SURE THE CELLS FACE FORWARD.

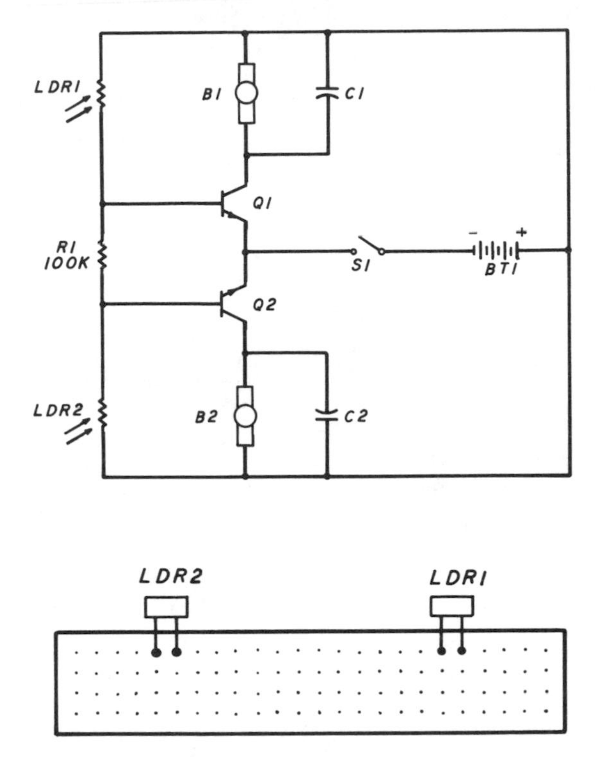

B) Insert two transistors (#18) in the middle of the first row of the board (between the two photocells). These are labelled Q1 and Q2 on the schematic. If you use Radio Shack #276-2014 or #276-2009 transistors, the transistor's flat side will face toward the back of the protoboard. If you use different NPN transistors or PNP transistors, make sure the leads are in the same order (emitter-base-collector) as the example.

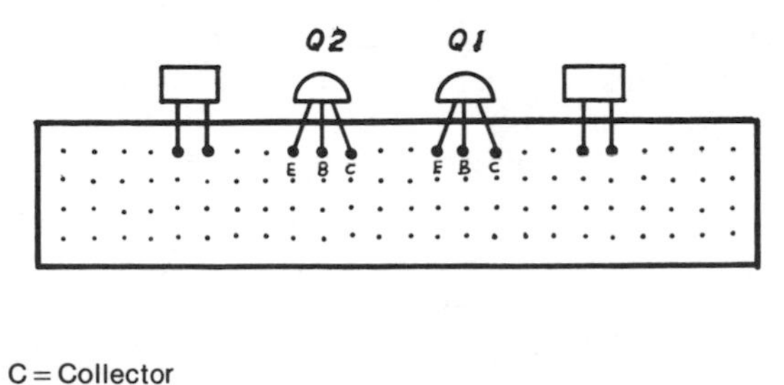

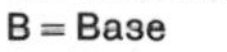
C = Collector
B = Base
E = Emitter

C) The photocells are wired in series with a resistor across the two inside leads. This resistor is designed to limit the amount of current flowing into the base of the transistors to an acceptable level. Take a 100K resistor (#19). Insert one end of the resistor into the hole in the protoboard's second row directly behind the left photocell's (LDR2) inside lead. Insert the other end into the hole directly behind the right cell's (LDR1) inside lead. If the resistors leads are too long and keep the resistor from lying close to protoboard, trim the leads slightly.

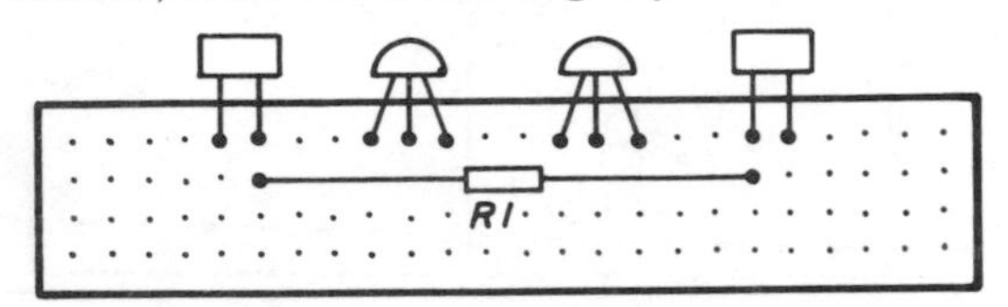

D) Check the schematic. See how the inside lead of each photocell comes into the base lead of a transistor. Cut two 3/4 - 1" lengths of wire (#20). Strip 1/4" of insulation off each end.

—Insert one end of the first wire into the hole in the protoboard's third row directly behind LDR2's inside (right) lead. Insert the other end into the hole in the third row directly behind Q2's base (middle) lead.

—Insert one end of the second wire into the hole in the protoboard's third row directly behind LDR1's inside (left) lead. Insert the other end into the hole in the third row directly behind Q1's base (middle) lead.

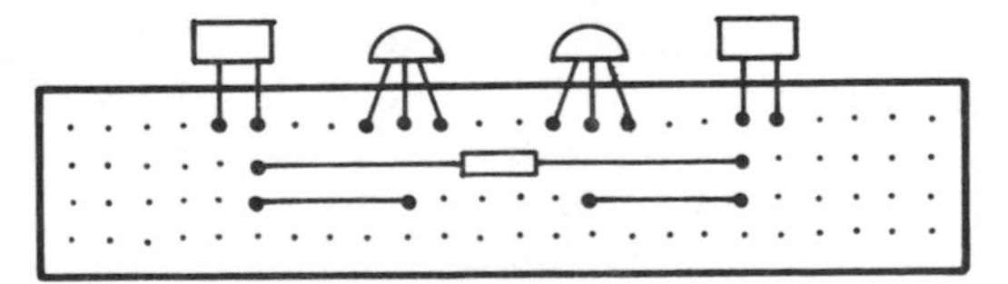

E) Cut a 1" length of wire, strip 1/4" of insulation off each end. Insert one end of the wire in the hole in the protoboard's fourth row directly behind the Q2's emitter lead. Insert the other end in the hole in the fourth row directly behind Q1's emitter lead.

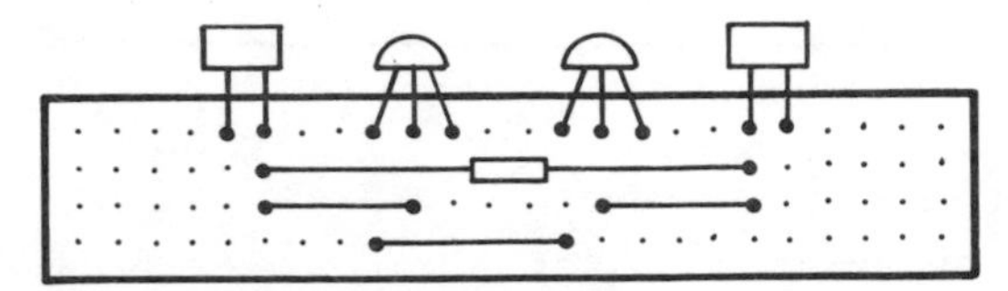

A WORD ABOUT PROTOBOARDS

You may want to experiment with the protoboard before wiring the Moth. Here is how the protoboard works. Orient the board in front of you so that the length runs right and left. Holes that run the length of the protoboard (the rows) are not connected to one another. Holes that run the width of the board (the columns) are connected. In other words, when leads are plugged into holes in the same column, they are connected just as if you had soldered the two leads together. Leads all lined up in a row across a protoboard are not connected.

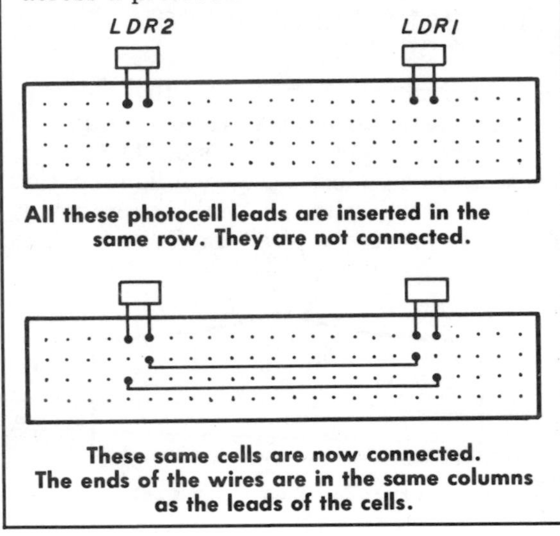

All these photocell leads are inserted in the same row. They are not connected.

These same cells are now connected. The ends of the wires are in the same columns as the leads of the cells.

F) Connect the two outside leads of the photocells. Cut a 2" piece of wire, strip 1/4" of insulation off each end. Insert one end of the wire into the hole in the fourth row directly behind LDR2's outside (left) lead. Insert the other end of the wire into the hole in the fourth row directly behind LDR1's outside (right) lead.

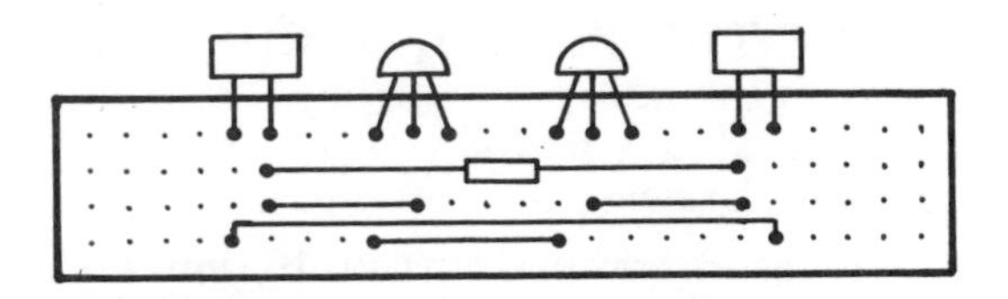

AT THIS POINT, THE PHOTOCELLS ARE WIRED IN SERIES WITH A

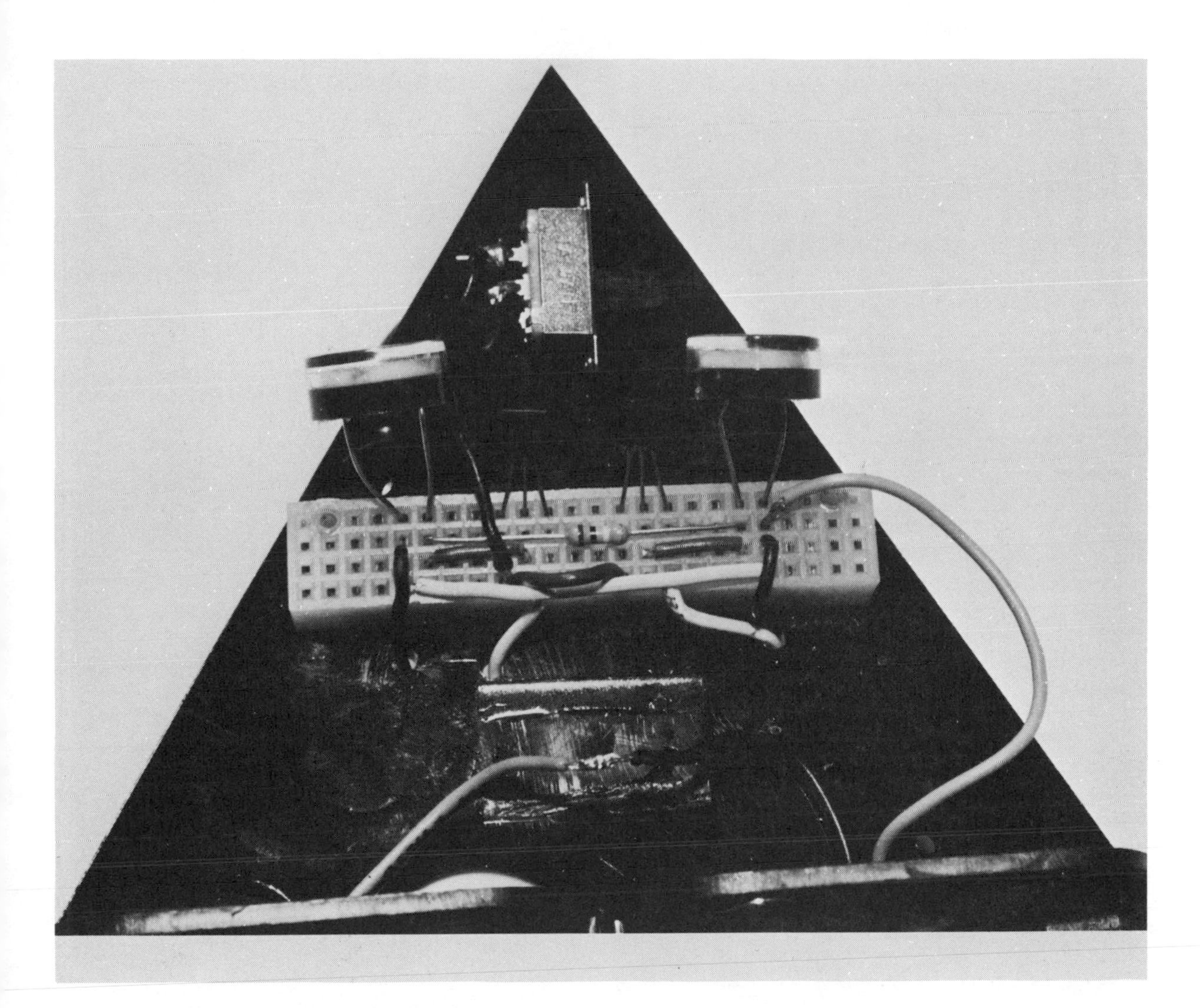

RESISTOR ACROSS THE INSIDE LEADS AND A WIRE ACROSS THE OUTSIDE LEADS. THE TRANSISTORS' EMITTER LEADS ARE CONNECTED, AND THE PHOTOCELLS ARE WIRED TO THE TRANSISTORS' BASE LEADS.

10. CONNECT THE MOTORS TO THE TRANSISTORS

One lead on each motor must be connected to the collector lead on each transistor. We are going to be tricky here. Instead of wiring the right motor to the right transistor and the left motor to the left transistor, we are going to cross the wires. This way, when you stand in front of the Moth and shine a flashlight into the right photocell, the Moth will turn right. Think about this—for the Moth to turn right, the left wheel must drive it to the right, and vice versa.

A) Now for trial and error: Your motor's polarity is regulated by the battery. Hook the motor leads to a battery one way and the motor shaft turns clockwise. Reverse the leads and it spins counter-clockwise. You could check the polarity by hooking the motor to a VOM and battery. But in the time it takes, you could just hook up the motors on the protoboard, test the system, and reverse the leads on a motor if the wheel spins the wrong direction.

B) With that in mind, take a lead from the *right* motor (B1 on the schematic), strip off 1/4" of insulation, and insert it in the hole in the fourth row directly behind the *left* transistor's (Q2) collector lead.

C) Take a lead from the *left* motor (B2), strip off 1/4" of insulation, and insert it in the hole in the fourth row directly behind the *right* transistor's (Q1) collector lead.

D) To keep things neat, we suggest that you drill two 5/32" holes in the platform. Locate them directly behind the protoboard, spaced about 1/2" either side of center. Slip the motor leads up through these holes before inserting them into the protoboard.

E) If any motor lead is too short, splice on a little extra wire. Cut the necessary length, strip off 1/4" of insulation. Twist one end into the short motor lead. Secure the joint with a little solder.

11) CONNECT MOTORS TO PHOTOCELLS

The remaining lead on each motor is wired to the outside lead of each photocell.

A) Take the remaining lead on the *right* motor (B1), strip off 1/4" of insulation, and slip through the same hole in the platform as the other B1 lead. Insert end in the protoboard's third row directly behind the outside lead of the *left* photocell (LDR2).

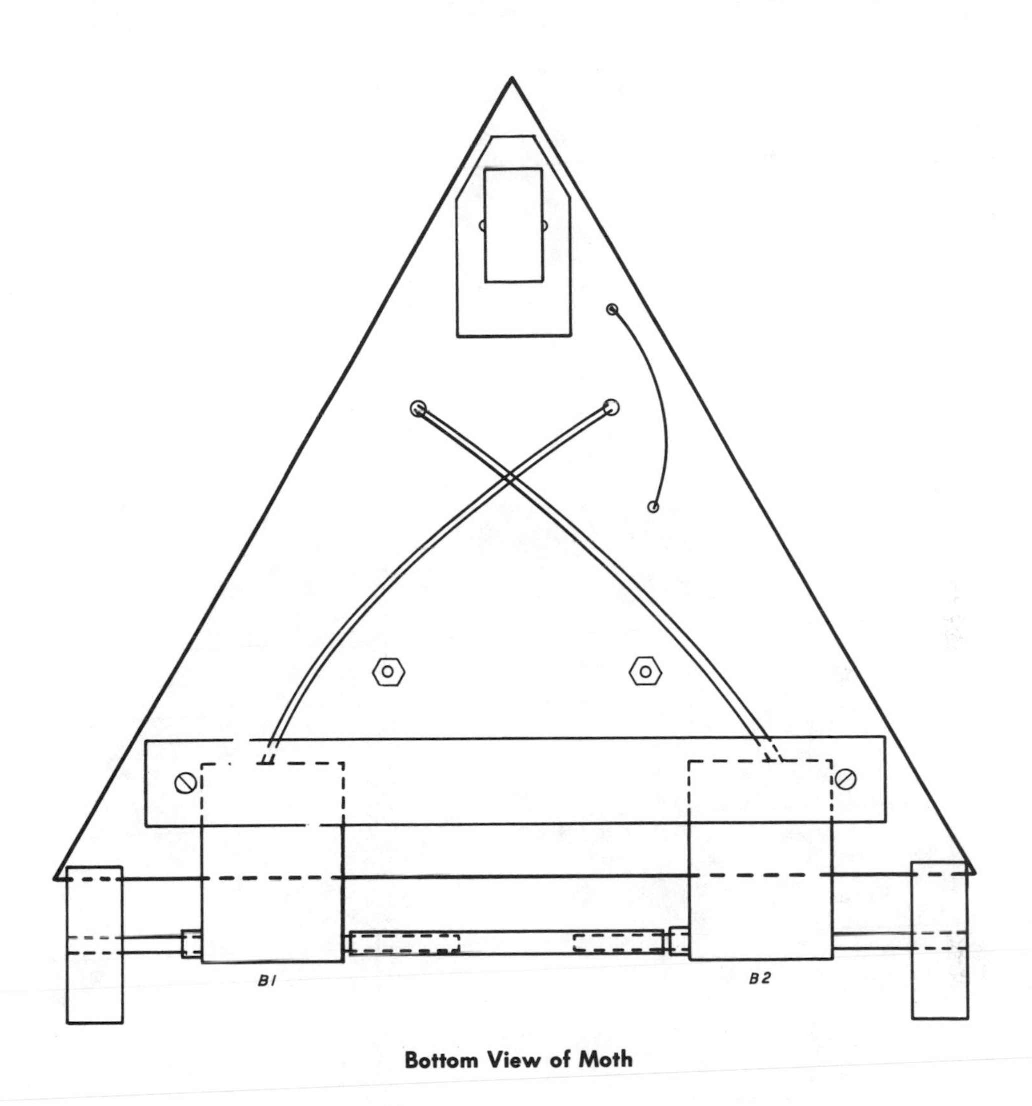

Bottom View of Moth

B) Take the remaining lead on the *left* motor (B2), strip off 1/4" of insulation, and slip through the same hole in the platform as the other B2 lead. Insert end in the protoboard's third row directly behind the outside lead of the *right* photocell (LDR1).

12. CONNECT BATTERY LEADS TO THE CIRCUIT

A) Insert the positive (red) lead from the battery holder (#4) into the protoboard's second row directly behind the outside lead of the right photocell (LDR1). The lead can be connected to the outside lead on either cell (since the two cells are wired in series), but LDR1 is closer to the positive lead on the battery holder. The lead from the battery holder is not solid core wire. If you have trouble inserting it in the protoboard, twist the end tightly and coat with a little solder.

B) Connect the negative (black) lead on the battery holder to the switch (#21). Neatness counts, so drill a 7/64" hole in the platform just in front of the left battery holder and another one in front of the protoboard. Avoid drilling through the base of the caster! Run the battery lead down through the first hole and back up through the second.

C) Connect the battery lead to the switch. There are two leads on the switch. Solder the battery lead to the switch lead

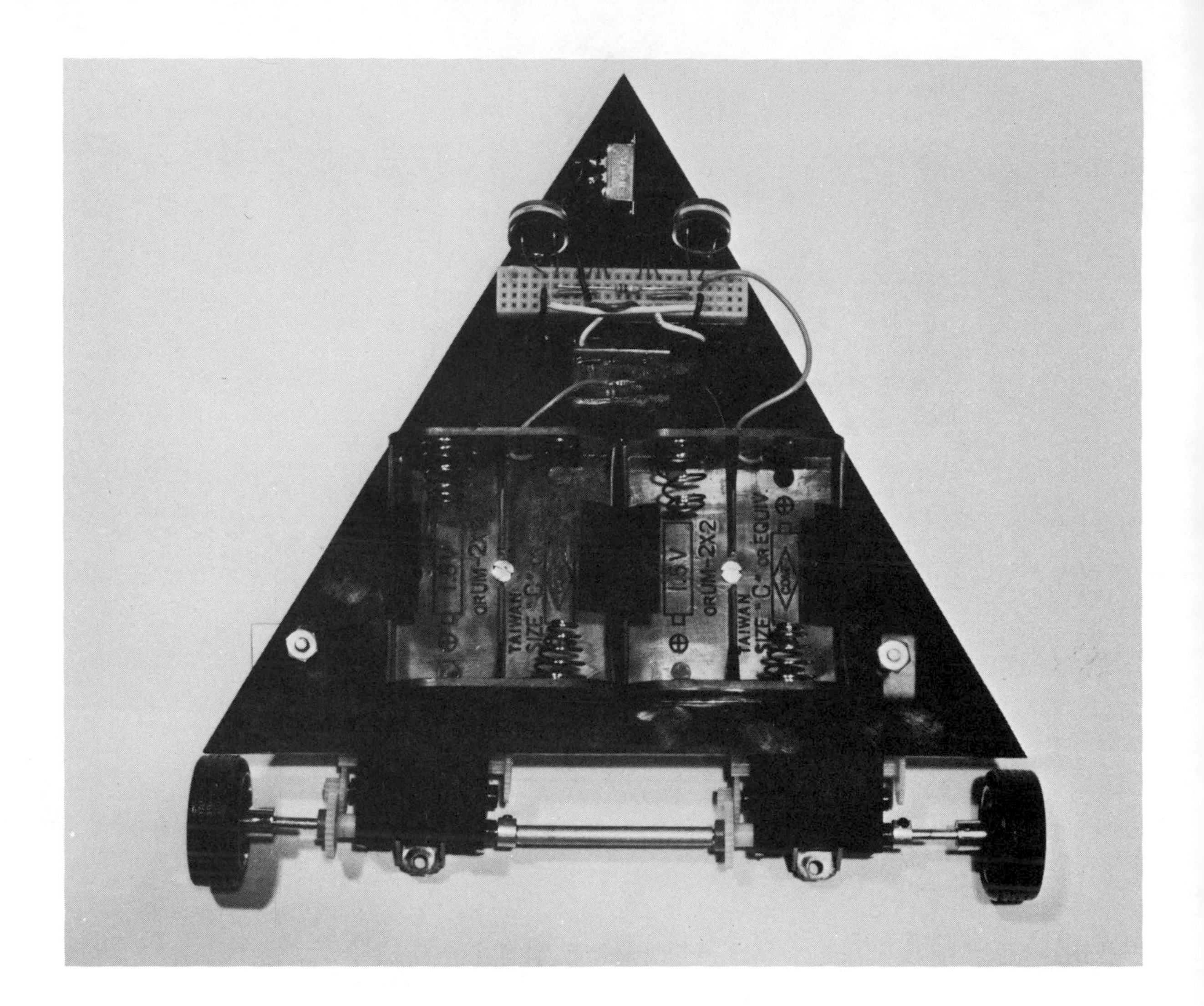

that is closest to the edge.

13. CONNECT SWITCH TO THE PROTOBOARD

A) Cut a 3" piece of wire, strip 1/4" of insulation off each end. Solder one end of the wire to the remaining (inside) lead on the switch.

B) Insert the other end of the wire into the protoboard's first row directly behind either transistors' emitter lead. Don't let the exposed end of this wire touch the resistor lead. You'll short the circuit, and the Moth won't run.

14. INSERT BATTERIES AND GO

A) Insert batteries (#22), taking care to observe the polarity as indicated on the battery holders. Switch on Moth. When you direct a flashlight at the photocells, the wheels should spin.

B) If either wheel spins in the wrong direction, turn off Moth and reverse leads on the motor that is spinning in the wrong direction.

15. GLUE DOWN THE SWITCH

14) Glue switch to the front of the Moth. Position the switch so that the "on" position

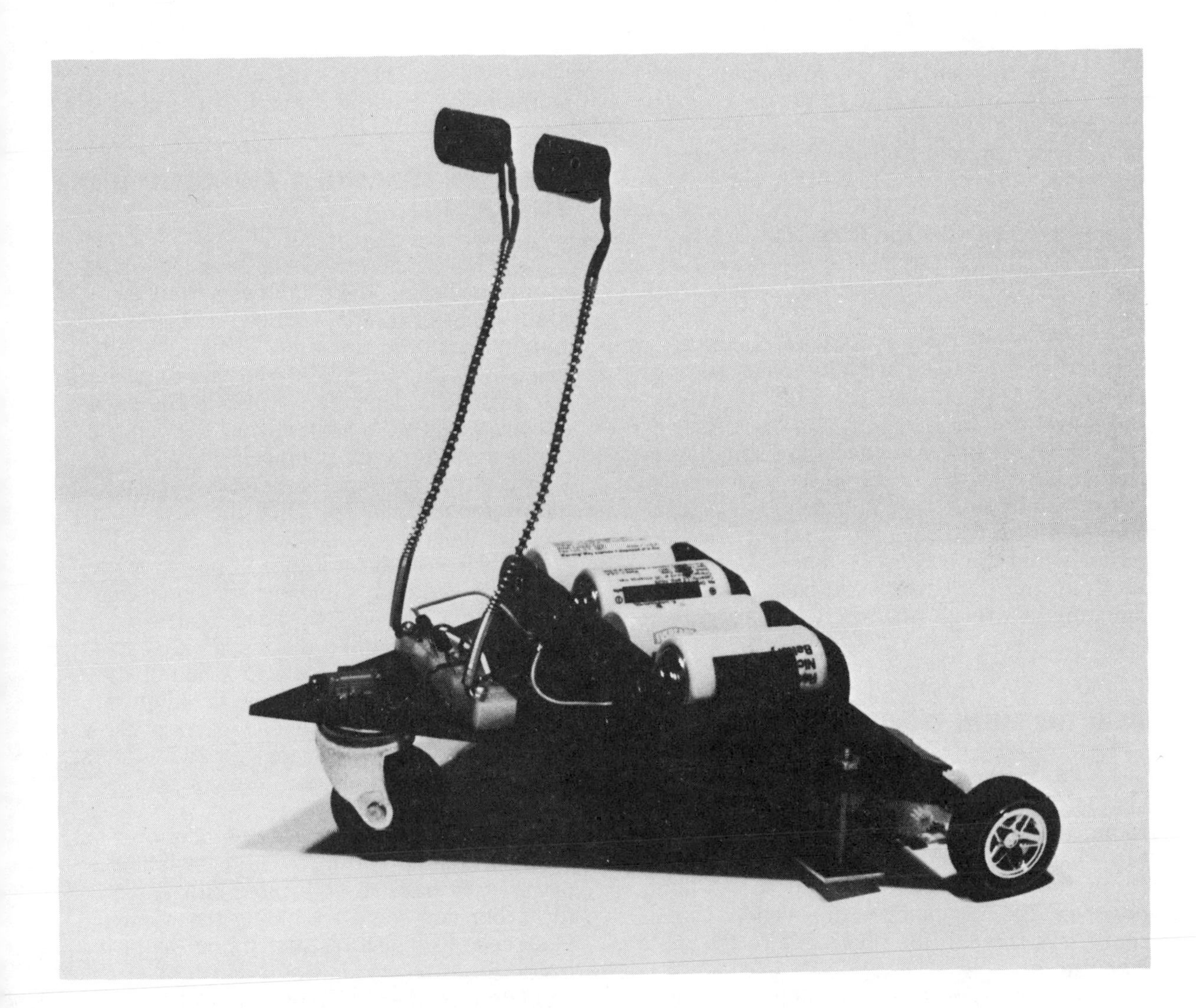

is toward the front of the Moth and "off" is toward the back. Cut a 2" length of plastic stick (#23) and glue to the lever on the switch. One end of the stick must protrude beyond the front end of the Moth, creating a "dead man's switch." If the Moth bangs into something, the stick will hit the object first and turn off the Moth. To soften the blow even more, put a piece of foam rubber (#24) on the end of the stick.

CHECKLIST FOR TROUBLESHOOTING THE MOTH

If your Moth does not work when you turn it on, don't worry. You may have to go through a little debugging process the way you do when writing a computer program.

You should first make certain that the batteries are in and connected properly. Are the nicads charged? Did you turn on the switch? If the answer is yes to the above questions, examine the wiring carefully. Check the schematic against your Moth. Is there a complete circuit for the electricity to flow through? Are all the components connected correctly? Are any wires touching that should not be touching? This could short out the system. Check the 100K resistor particularly closely. Its leads are not insulated.

Don't let this process frustrate you. Make a game out of troubleshooting the

Moth. Later, when you are developing your own robots, you will have to devise a checklist of your own.

Assuming you said yes to the above questions, let's check some other possibilities:

1. If you turn on the switch and nothing happens, make sure the transistors are in properly. The collectors of NPN transistors are connected to the motors. (See steps 10B and 10C)

2. Perhaps there is a bad connection somewhere and the electricity is not getting through the circuit. Use the VOM to check the current at various points in the circuit.

3. If one or both wheels are spinning in the wrong direction, reverse the leads coming out of the motors. (See step 10)

4. If the batteries are smoking you have either wired the battery holders (#4) incorrectly or put the batteries in backwards. The smoking is a result of reversed polarity.

HOW THE MOTH WORKS

The concept behind the Moth is very simple. When you turn on the robot in a dark room, the photocells have a high resistance rate that blocks the flow of electricity. The Moth does not move. When exposed to light, however, the resistance is diminished. Current flows through the photocells to the transistors.

The transistors act like amplifiers and increase the amount of current. The collectors send this boosted signal directly to the motors, which drive the Moth.

The schematic does not show it, but each motor is connected to the opposite transistor. The reason for this is that when you shine a flashlight at the right cell, you instinctively expect the Moth to move to the right (or toward the light). But it is actually the left wheel that pushes the Moth to the right, and vice versa. Because the crossed wires are not actually necessary to make the Moth run, we opted for simplicity in the schematic and did not cross the wires.

You may notice too that the brighter the light source, the faster the Moth moves. This is because a brighter light lowers the photocells' resistance even more and permits more current to flow through the circuit.

IDEAS FOR IMPROVING AND CUSTOMIZING YOUR MOTH

1. Make the photocells look like antennae by mounting them up away from the Moth's platform. We salvaged a couple of 6'' springs from two toy tanks. Then, we cut four lengths of wire, soldered one to each of the photocells' leads, and covered the exposed wire leads with heat shrink. We ran the two leads from each photocell down the length of the springs, screwed the springs to the protoboard, and plugged the wires into the protoboard.

2. Use the basic Moth to power models and toys. We built a plastic car model kit. The kit was powered by a DC gearhead motor. We redesigned the wiring a bit and mounted a photocell in place of one of the headlights. When we shined a flashlight at the car's front end it moved. A larger car model would have had room for two motors, and we could have steered the car by flashlight.

3. Mount a plastic robot model on a clear, plexiglas platform. The wheels and motors will have to be located on the platform, but you could try hiding the wiring, photocells, transistors, switch, and batteries inside the robot. This way you could animate one of your favorite models.

4. Decorate your robot with flashing LEDs. Start by inserting a couple of LEDs in series between the photocells and the transistors. Remember, LEDs have polarity like any diode. Also, you will have to insert resistors to keep the LEDs from burning out. Use Ohm's Law to determine what size resistor to use.

COMPUTERIZING THE MOTH

Although the basic Moth is a fairly unsophisticated robot, it represents a good starting point for a new robot builder. You are familiarizing yourself with electronic components and understanding their functions in a circuit. You have used this knowledge to construct a remote-controlled, sensory robot.

But one of the criteria for a robot is having a computer for a "brain" (central processor). Let's now examine how you might connect a personal computer to your Moth. In addition to computerizing the Moth, you will gain the information you need to control other peripheral devices via the computer. Some of this preliminary information may be confusing at first. For many of you this is your first time working with the user ports and integrated circuits (ICs). Read everything carefully, but if you do not understand how the pin values were determined and the ICs were selected, don't worry. Just follow the steps for connecting the Moth to the computer; little by little you'll understand how it all works.

The first step in making any connection to your computer is to check your computer's technical manual for data on the I/O (input/output) ports. You need to know memory locations, address values, pin connections, and the preset state of those pins (high or low, which is the same as saying on or off). We have included instructions for interfacing the Moth to the Commodore 64, but the development process is essentially the same for any computer. The memory locations, address values, and pin connections will differ.

SHOPPING LIST

1) **Experimenter's Interface Board**—This is a homebrew protoboard and cable that you can use to interface the Commodore 64 to peripherals such as the moth. See building instructions on page 69.

2) **2 optocoupler ICs** (type TIL-111)—You will get at least 2 of these ICs in Radio Shack's package of assorted optocouplers #276-1654, $1.98.

3) **2 1/4-watt, 470 ohm resistors** (yellow-violet-brown)—Radio Shack #271-1317, $.39 for 5.

4) **2 1/4-watt, 100K ohm resistors** (brown-black-yellow)—Radio Shack #271-1347, $.39 for 5.

5) **2 1/4-watt, 3.3K ohm resistors** (orange-orange-red)—Radio Shack #271-1328, $.39 for 5.

6) **3-wire solid intercom cable** with color coated insulation—Radio Shack #278-370; 7' for $5.59.

CONNECTING THE MOTH TO THE COMMODORE 64

1) READ THE JOYSTICK

Since we want to control the Moth with a joystick, we need to determine the address values for the forward, left, and right positions on the joystick. The following program reads the value for each position:

```
10 PRINT "(SHIFT-CLR/HOME)":REM
   CLEAR SCREEN—YOU SHOULD SEE
   AN INVERSE HEART INSIDE THE
   QUOTATION MARKS
20 PRINT PEEK (56321):REM THIS IS THE
   ADDRESS OF JOYPORT B
30 GOTO 10
```

2) CALCULATE JOYSTICK VALUES

When we run the program, we get different values depending on the position of the joystick. The values are as follows:

Position	Value	Bit Value	Reduced Value
Center	255	off	0
Forward	254	2^0	1
Left	251	2^2	4
Right	247	2^3	8

The highest value is 255. In order to reduce these numbers to values of a single binary word bit, we subtract each position value from the center (off) position value of 255. We have provided the bit value conversion as well. For example, 255-247 (right) = 8 = 2^3 (bit value).

3) CONTROL THE I/O PORT

A) The joyport 1 is for our input to the computer. Output from the computer is through the user port (this is the 24-pin port on the back of the Commodore 64, not to be confused with the much longer cartridge expansion port).

B) To make the connection between the joyport and the user port, we need to locate the pins in the user port that correspond to the address values for the three joystick positions. Page 360 of the *Commodore 64 Programmer's Reference Guide* shows that *Pin E* equals 4 (PB2 = Port B 2^2), which corresponds to *left,* and *Pin F* equals 8 (PB3 = Port B 2^3), which corresponds to *right.*

C) To make the Moth go forward, we could use Pin C on the user port, which corresponds to the joystick forward value of 1. It is even simpler, however, just to turn on both the left and right motors simultaneously. All we do is add left (4) and right (8). A value of 12 will make the Moth go forward. You'll see how this is used in the short computer program to follow.

4) SELECT ICs

A) We discover that the pins are preset to high values, which means that they are on all the time. In digital logic, a high state (designated by a 1) is about +5VDC; a low state (designated by a 0) is approximately 0VDC (or off). To keep the Moth from running all the time, we must set the pins to low logic values. The easiest way to turn them off with a switch.

B) Two optocoupler ICs serve as our switches. The optocoupler we selected contains an infrared emitting diode (similar to an LED) and a photo transistor. The photo transistor works on the same principle as the ones in our Moth, with one exception. It is turned on by the light from the infrared emitting diode. We control the switch with input values from a joystick or a computer program.

5) CONNECT EXPERIMENTER'S INTERFACE BOARD

With the preliminaries out of the way, we can connect the computer to the Moth. To simplify the wiring process, we designed a special Experimenter's Interface Board (EIB). It's a protoboard that connects directly

ARE YOU STILL WITH US?

Let's review what we have done so far. We:

1) Wrote a program to PEEK memory and read the address values for the joystick positions (joysticks provide input to the computer);

2) Converted those values to binary bit values;

3) Located the pins in the user port that correspond to the address values of the two joystick positions (left and right);

4) Combined the values of left (4) and right (8) to get 12. This value turns on both motors simultaneously and lets the Moth go forward.

5) Used two optocouplers to act as switches to control the high/low output values from the user port pins.

to the Commodore 64's user port. To build the board, see page 69. Now, plug the Experimenter's Interface Board into your computer. Make sure your mark on the edge connector faces up (see EIB project).

6) INSERT ICs ON EXPERIMENTER'S INTERFACE BOARD

A) The first step in wiring the protoboard is to insert the two ICs. Notice the space running through the center of the protoboard. Straddle the pins on the ICs across this space. The space should be directly under the center of each chip. The small depression on each chip is located close to Pin 1. Use this depression to orient the ICs with Pins 1-3 facing toward the top of the Interface Board (closest to the jumper sockets). When inserting the chip, be sure the pins line up with the holes on the protoboard. In other words, a row of pins should all be in the same row of the protoboard. Gently push the ICs down so as not to bend any pins.

B) Looking down on the chip, the small depression will be in the upper righthand corner. Pin 1 is in the top right position. The pin values increase in a counter-clockwise direction.

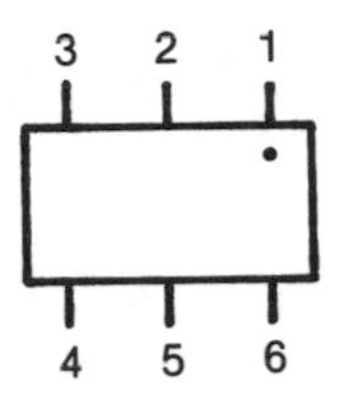

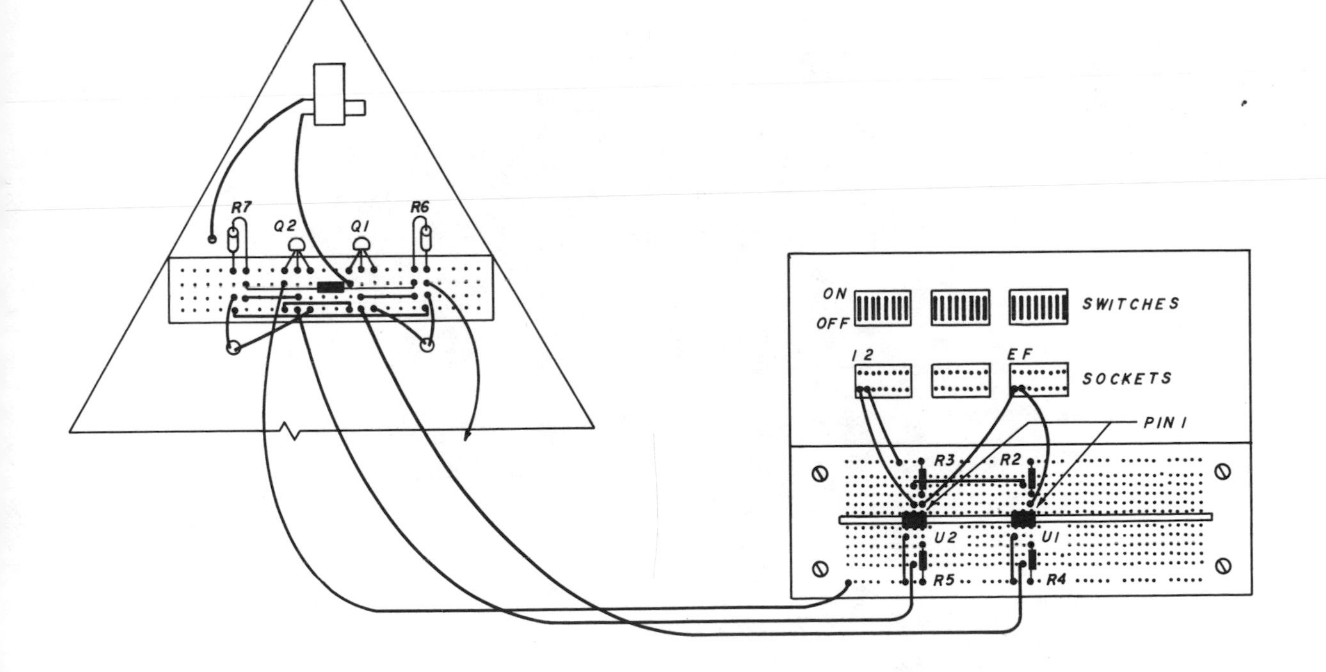

7) REDESIGN MOTH'S PROTOBOARD

Since you are replacing the photocells (LDR1 and LDR2) with the computer interface, remove LDR1 and LDR2 from the Moth's protoboard. Replace them with a couple of 3.3K ohm resistors. Put the resistor lead into the exact same holes as LDR1 and LDR2.

8) CONNECT PIN E

To make the Moth turn left, the joystick value is 4. This corresponds to Pin E in the user port.

A) Run a 2'' piece of wire from the jumper socket corresponding to Pin E on the EIB to the hole directly above Pin 1 on the left optocoupler (U2).

B) Take the intercom cable and cut it to whatever length you want. Since the cable connects the Moth to the EIB, the longer the cable the farther the Moth can travel. Carefully remove 2'' of the outer insulation and 1/4'' of insulation from each of the individual wires. Do this to both ends of the intercom cable.

C) Select one of the wires inside the intercom cable. Insert one end into the hole directly below Pin 5 of the left optocoupler (U2). Insert the other end of the same wire (same color) into the hole in the fourth row of the Moth's protoboard directly behind the left transistor's (Q2) base lead.

9) CONNECT PIN F

To make the Moth turn right, the joystick value is 8. This corresponds to Pin F in the user port.

A) Run a 2'' piece of wire from the jumper socket corresponding to Pin F on the EIB to the hole directly above Pin 1 on the right optocoupler (U1).

B) Select another of the wires inside the intercom cable. Insert one end into the hole directly below Pin 5 of the right optocoupler

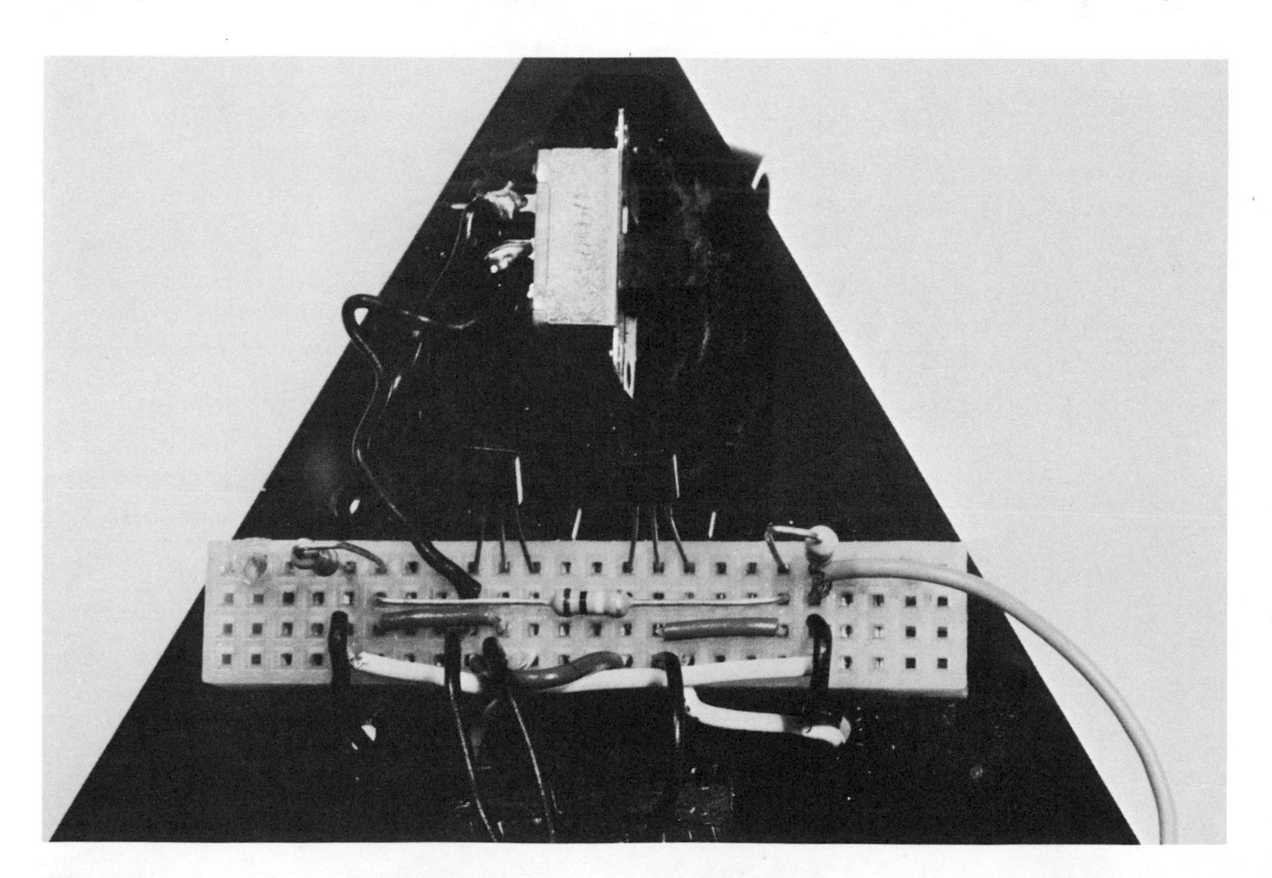

(U1). Insert the other end of the same wire (same color) into the hole in the fourth row of the Moth's protoboard directly behind the right transistor's (Q1) base lead.

10) CONNECT OPTOCOUPLERS' EMITTER AND BASE LEADS

A) Cut a 1'' piece of wire and strip 1/4'' of insulation off each end. Insert one end into the hole directly below Pin 4 on the left optocoupler (U2). Insert the other end of the wire into a hole in the bottom bus (the single row of holes along the bottom of the EIB).

B) Take a 100K ohms resistor. Insert one end in the hole directly below Pin 6 on U2. Insert the other end into a hole in the bottom bus. Trim the resistor's leads so that it rests right on the EIB.

C) Repeat steps 10A and 10B for the right optocoupler (U1).

11) CONNECT COMPUTER'S POWER SUPPLY AND GROUND TO ICs

A) Run a 3'' piece of wire from Pin 1's jumper socket to the hole directly above Pin 2 on the left optocoupler (U2). Pin 1 on the user port is the computer's ground connection.

B) Run a 1-1/2'' piece of wire from Pin 2 on the left optocoupler (U2) to Pin 2 on the right optocoupler (U1). This connects U1 to ground.

C) Run a 1-1/2'' piece of wire from Pin 2's jumper socket to the top bus (the single row of holes at the very top of the protoboard). Pin 2 on the user port is the computer's 5VDC power supply.

D) Connect the 5VDC power supply to the ICs. Run a 470 ohms resistor from the hole directly above Pin 1 on the left op-

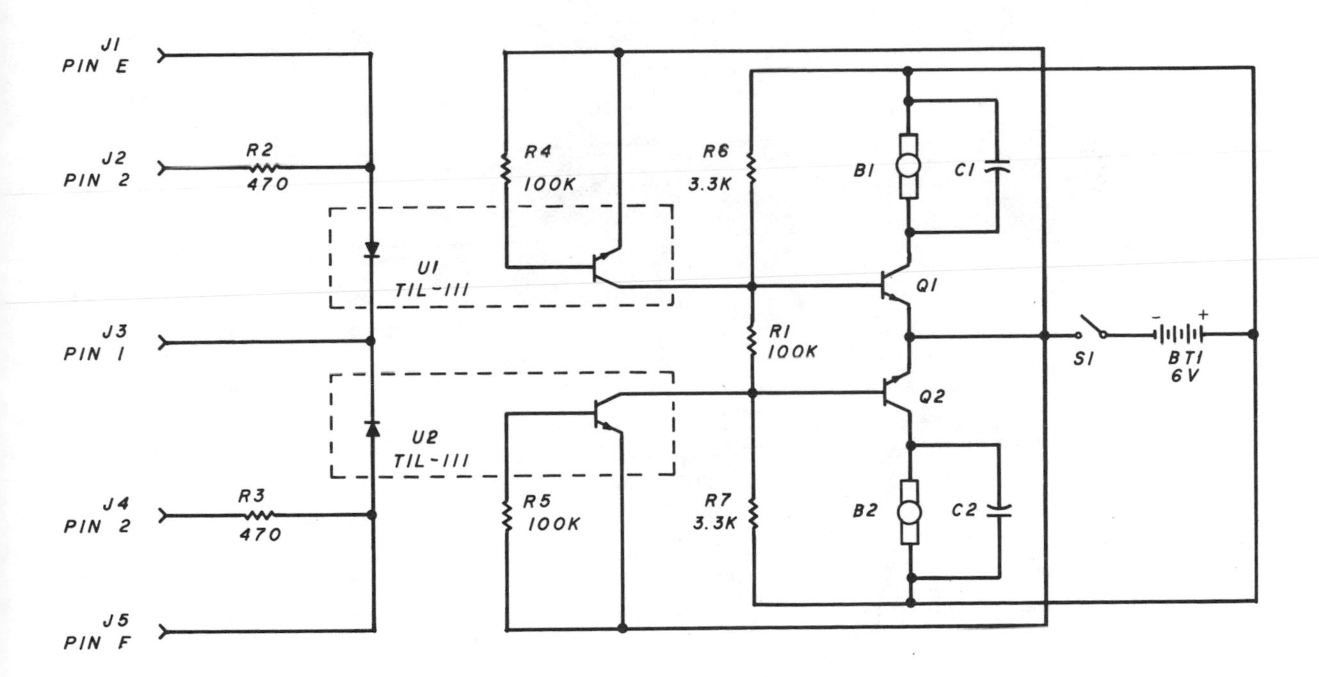

tocoupler to the top bus. The resistor reduces the voltage coming into the IC to a level that the infrared emitting diode can handle. The specs on the IC are printed on the back of the Radio Shack package. The diode can handle only 1.7 volts.

E) Run a second 470 ohms resister from the hole directly above Pin 1 on the right optocoupler to the toy bus.

12) COMPLETE THE CIRCUIT

A) Take the remaining wire in the intercom cable and insert one end into the EIB's bottom bus.

B) Insert the other end into the hole in the third row of the Moth's protoboard directly behind the left transistor's (Q2) emitter lead.

13) TURN ON DIP SWITCHES

To open up communication between the computer and the EIB, flip on the DIP switches for Pin 1, Pin 2, Pin E, and Pin F. If you just want to count the switches from left to right, they would be switches 1, 2, 17, and 18. Now you are ready to control your Moth via the Commodore 64's joystick. Just enter the following program, plug the joystick into port 1 and have fun.

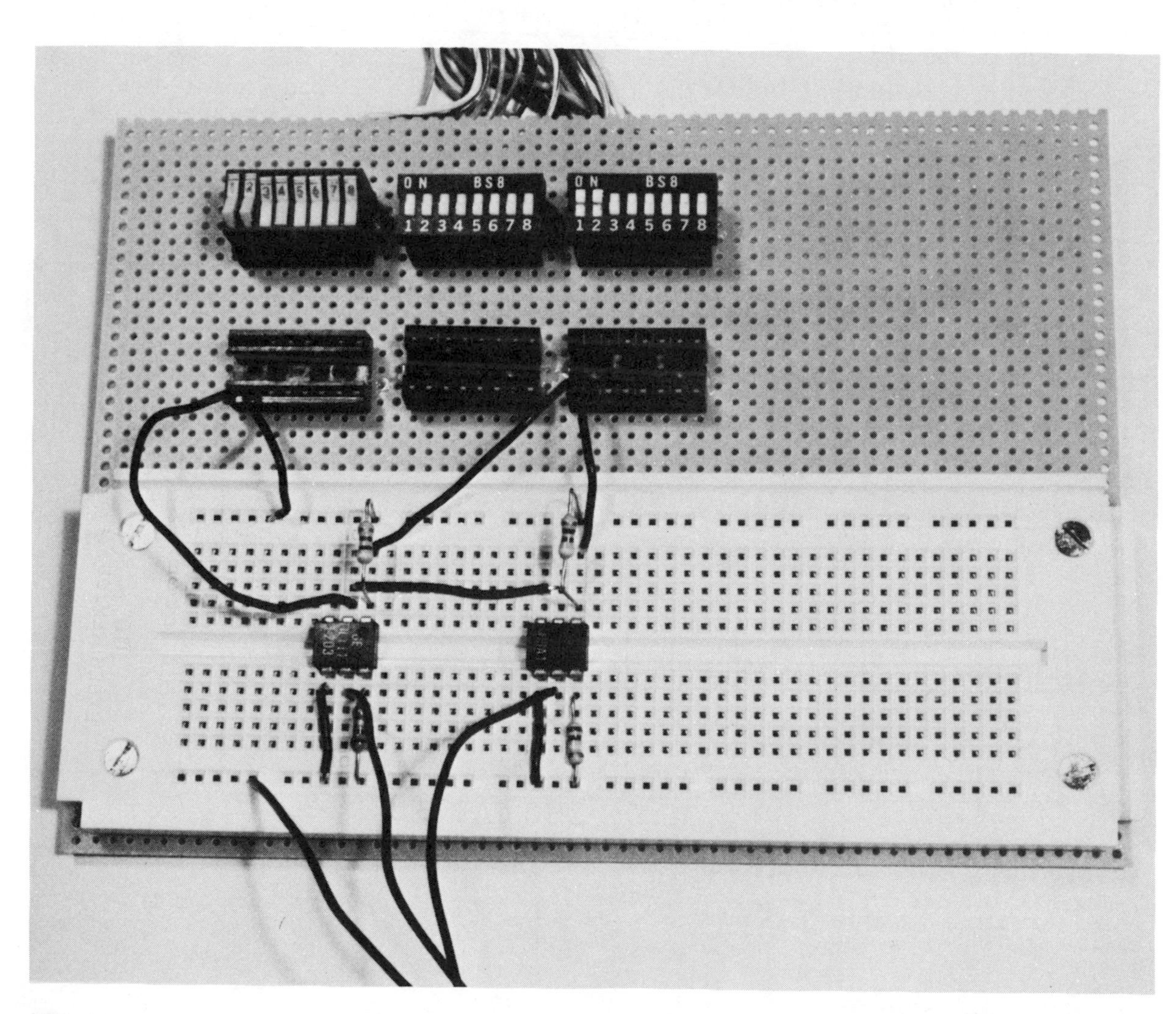

```
10 REM CONVERT JOYSTICK INPUTS INTO USER
   PORT OUTPUTS
20 PRINT "(SHIFT-CLR/HOME)":REM CLEAR THE
   SCREEN
30 I = PEEK(56321):REM PORT B INPUT VALUE
40 X = 255 - I:REM CONVERTS BIT VALUE FROM
   MSB TO LSB
50 PRINT X
60 F X=1 THEN X=12:REM 1 IS THE JOYSTICK
   VALUE FOR FORWARD. WE ARE CONVERTING
   IT TO 12 (4+8), WHICH TURNS ON BOTH
   MOTORS
70 POKE 56579,X:REM OUTPUT TO ROBOT
80 GOTO 20
```

14) PROGRAM YOUR MOTH

If you don't want to control the Moth with a joystick, here's a sample of the kinds of programs you can write. We have used random numbers to make the Moth dance. Can you write a program to make the Moth trace out a square?

```
10 REM THE DANCING MOTH
20 PRINT "(SHIFT-CLR/HOME)":REM CLEAR
   SCREEN
30 POKE 56579,0:REM TURN OUT VALUES OFF
40 X = INT(RND(0)*5)+4:REM SELECT WHOLE
   NUMBERS 4-8
50 IF X = 5 THEN X=1:REM THIS IS AN UNUSED
   VALUE; MOTH STANDS STILL
60 IF X = 7 THEN 40:REM SELECT ANOTHER
   VALUE
70 IF X = 6 THEN X=12:REM A VALUE OF 12
   TURNS ON BOTH MOTORS
80 Y = INT(RND(0)*100):REM RANDOM NUMBER
   FOR DELAY LOOP
90 FOR I = 1 TO Y:REM USE DELAY LOOP TO
   CONTROL TIMING OF MOTH'S ACTIONS
100 POKE 56579,X:REM OUTPUT TO ROBOT
110 PRINT X,Y,I:REM SEE THE OUTPUT ON THE
   MONITOR
120 NEXT I
130 GOTO 20
```

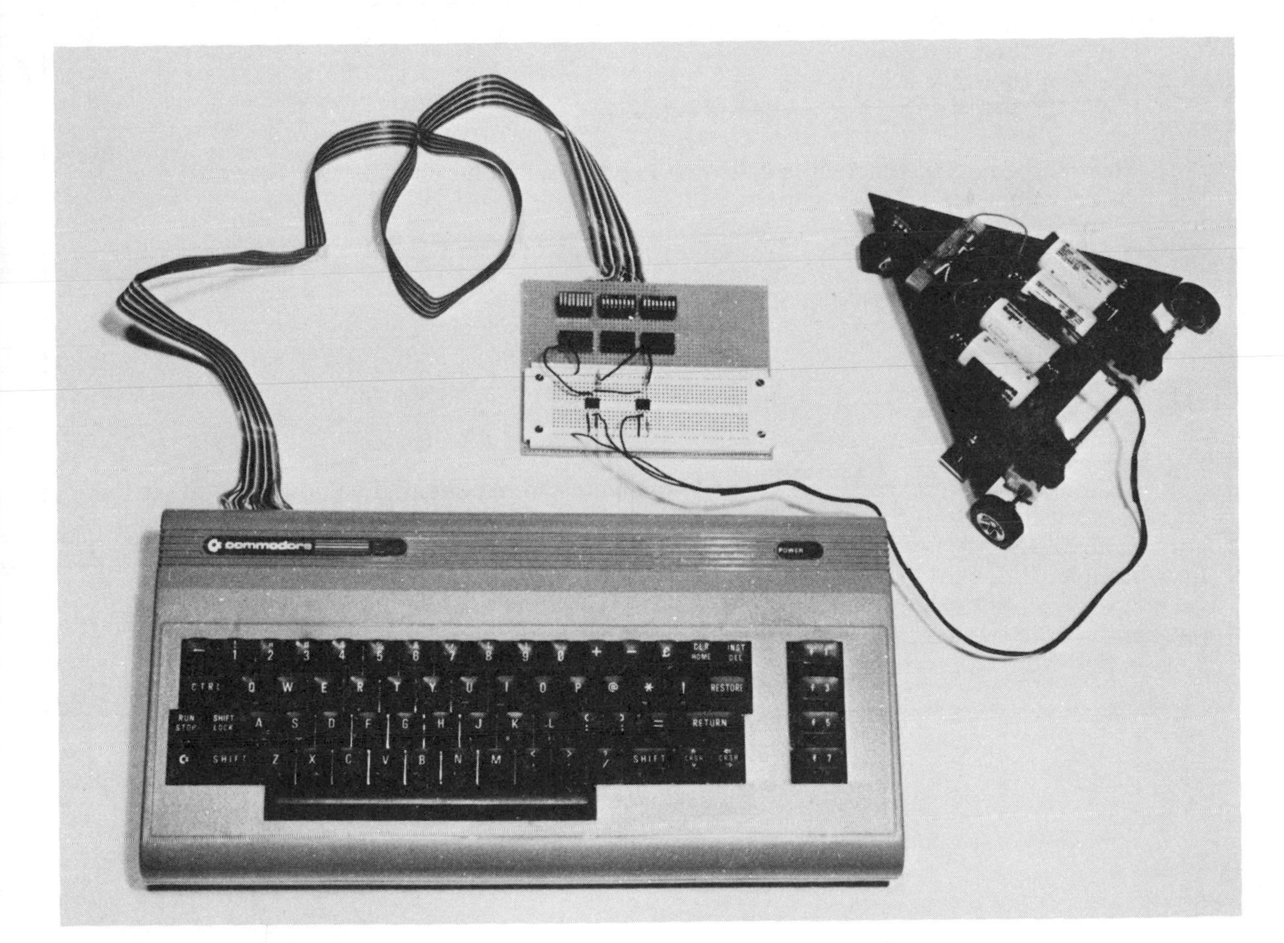

WHERE DOES THE VALUE OF 255 COME FROM?

In case you are wondering where these values came from, the Commodore 64 uses 8-bit words. A bit is a single unit block that contains the value of one or zero. A word is a series of individual bits. Therefore, an 8-bit word is a series of 8 bits following one after another.

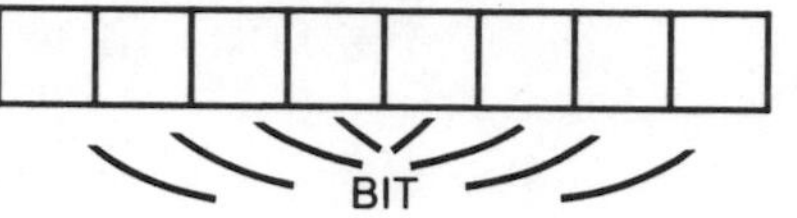

In a word, each bit is given a hierarchial value. This hierarchy (designated n) will determine the value of the word. Start with the right most bit, called the Least Significant Bit (LSB), and read to the left. This is just the opposite of reading a book. The first bit has the hierarchial value 0, the second bit has the value 1, the third bit 2, fourth bit 3, . . . eighth bit 7. This last bit is called the Most Significant Bit (MSB).

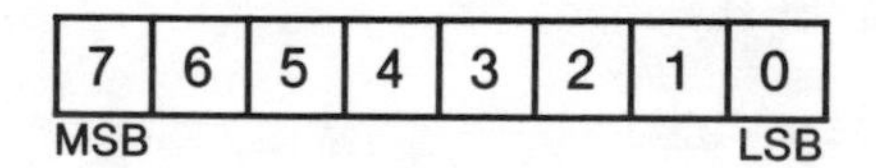

Remember that a bit has only two possible states or values (0 or 1). These two values and the bit hierarchy determine the numerical value of a bit. Continue multiplying 2 times itself to the number of the hierarchy. For example:

If n = 2 then 2 x 2 = 4
If n = 4 then 2 x 2 x 2 x 2 = 16
If n = 7 then 2 x 2 x 2 x 2 x 2 x 2 x 2 = 128

This mathematical function is called an exponential. The chart below shows all the possible bit values for an 8-bit word:

Hierarchy (n)	Exponential (2^n)
0	$2^0 = 1$
1	$2^1 = 2$
2	$2^2 = 4$
3	$2^3 = 8$
4	$2^4 = 16$
5	$2^5 = 32$
6	$2^6 = 64$
7	$2^7 = 128$

If the bit of a word has been set to 1, calculate the exponential value of the bit. If the bit of a word has been set to 0, ignore the bit and move to the next. To determine the numerical value of a word, add up all the exponential values from all the bits containing a 1. For example, the word 00011011 has a 1 in hierarchy bits 0, 1, 3, and 4. Using our table to get the exponential values, add 1 + 2 + 8 + 16 = 27. Therefore, the word 00011011 has the binary word value of 27. If all the bits are high, then the word 11111111 equals 255. This is the same value returned when the joystick is in the center position.

BUILDING THE EXPERIMENTER'S INTERFACE BOARD

The Experimenter's Interface Board (EIB) is a protoboard that is wired directly to the computer through one of the external ports. You can build an EIB to interface with any computer. This particular project, however, describes the EIB for the Commodore 64. It connects to the computer through the 24-pin user port.

You will notice that 24 DIP switches run across the top of the EIB. These allow you to control the 24 pins in the user port. You can turn a pin on or off just by flipping the appropriate DIP switch.

In this book, we use the EIB to connect the Commodore 64 to the Moth. We're sure that you will think of other ways to use your EIB.

EIB SHOPPING LIST

1) Modular IC protoboard (breadboard)—2-1/8" x 6" Radio Shack #276-174; $11.95.

2) IC-spacing perforated board—4-1/2" x 6" Radio Shack #276-1394; $1.89.

3) Package each of #4-40 x 1/2" **flat head steel machine screws;** #4-40 x 1/4" x 3/32" steel hex nuts; #4 steel split lock washers—When assembling, put washer on just before nut. This helps keep screws from loosening.

4) 6 **16-pin wrap DIP sockets**—Radio Shack #276-1994; 2 for $1.39.

5) 3-foot, **24-position ribbon cable**—Radio Shack does not carry this size, but you should be able to get it at any good electronics store. Be sure and get the rainbow-colored cable; $3.

6) 24-position card-edge socket—Radio Shack does not carry this size, but you should be able to get one at any good electronics store; $1.95.

7) Wire wrapping tool—Radio Shack #276-1570; $5.95.

8) 30-gauge Kynar wire—This is wrapping wire. Radio Shack #278-502, 50 feet for 2.39.

9) 3 **8-position SPST DIP switches**—Radio Shack #275-1301; $1.99 each.

EXPERIMENTER'S INTERFACE BOARD FOR THE COMMODORE 64

1. MOUNT PROTOBOARD

A) Position and center the protoboard (#1) at the bottom of the perforated board (#2). See photograph for top view of the Interface Board.

B) Align each screw hole on the protoboard with a hole on the perforated board. Mark these holes on the back of the perforated board. Use a 1/8" drill bit to drill out the marked holes.

C) Reposition the protoboard and secure each corner with a flat head machine screw, split lock washer, and hex nut (#3). The nuts should be located on the back side of the perforated board.

2. MOUNT DIP SWITCH SOCKETS

A) Take three of the sockets (#4); place them on the perforated board above the protoboard. Leave enough space for a second row of sockets under the first.

B) Align the three sockets across the perforated board, leaving at least two holes between socket edges. Turn the board over and spread the socket leads apart slightly. This will hold them in place until you attach wires.

3. MOUNT JUMPER WIRE SOCKETS

Take three more sockets (#4); place them between the first row and the protoboard. Leave a little space between the two rows. Carefully align each socket with the one above. Again, spread the socket leads slightly.

4. CONNECT RIBBON CABLE TO CARD-EDGE SOCKET

A) Take the card-edge socket (#6) and examine the solder connector edge (side that has all the pins protruding). Locate the row of pins marked 1 through 12. This is the top of the connector. Mark the top with a magic marker or pencil so that you can easily recognize it. If your card-edge socket is unmarked, just designate one of the rows as pins 1 through 12 and mark that side as the top. (Until you start connecting the cable (#5), the two rows are interchangable.)

B) Locate Pin 1. As you look at the solder connector edge, the top, left pin is Pin 1. The top, right pin is Pin 12. Similarly, the bottom, left pin is Pin A, and the bottom, right pin is Pin N. NOTE: THERE IS NO PIN G OR PIN I ON THE BOTTOM ROW. You can see a picture of the pins and a

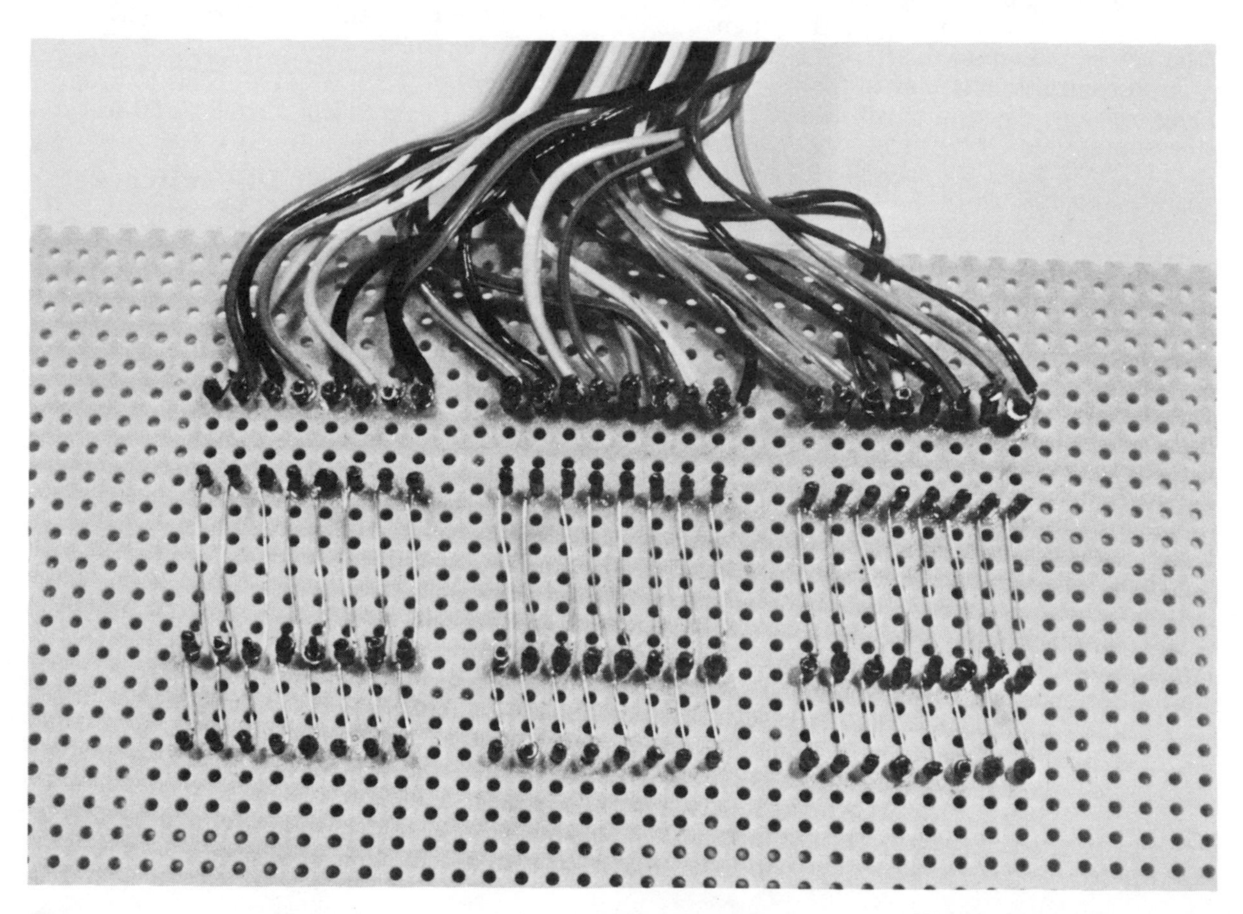

1	2	3	4	5	6	7	8	9	10	11	12
A	B	C	D	E	F	H	J	K	L	M	N

description on page 397 of the *Commodore 64 Programmer's Reference Guide.*

C) Peel back the individual wires in the cable about 2" from each end. The ends of the cable will look like fringe. This makes it easier to connect the individual wires to the solder pins. Strip about 1/4" of insulation off each of the wires at one end of the cable and about 1" of insulation off wires at the other end.

D) Now you can begin to solder the 24 wires to thc 24 pin connections (use the wires that have 1/4" of insulation removed). *The wiring is tricky, so read carefully:*

—Start with the first wire on the left edge of the cable. Connect it to *Pin 1.* Just slip it through the hole in the pin, twist the wire to make a good connection, and solder.

—Take the second wire from the left and solder it to *Pin A*

—Solder the third wire from the left to *Pin 2.*

—Continue this pattern with the remaining wires. NOTE: THE CONNECTION PATTERN IS TOP ROW, BOTTOM ROW, TOP ROW, ETC. Here is the wiring order:

1-A-2-B-3-C-4-D-5-E-6-F-7-H-8-J-9-K-10-L-11-M-12-N

5. CONNECT THE RIBBON CABLE TO THE DIP SWITCH SOCKETS

A) Pay close attention here: When you look at the Interface Board from the *top,* it makes sense to designate the leftmost socket

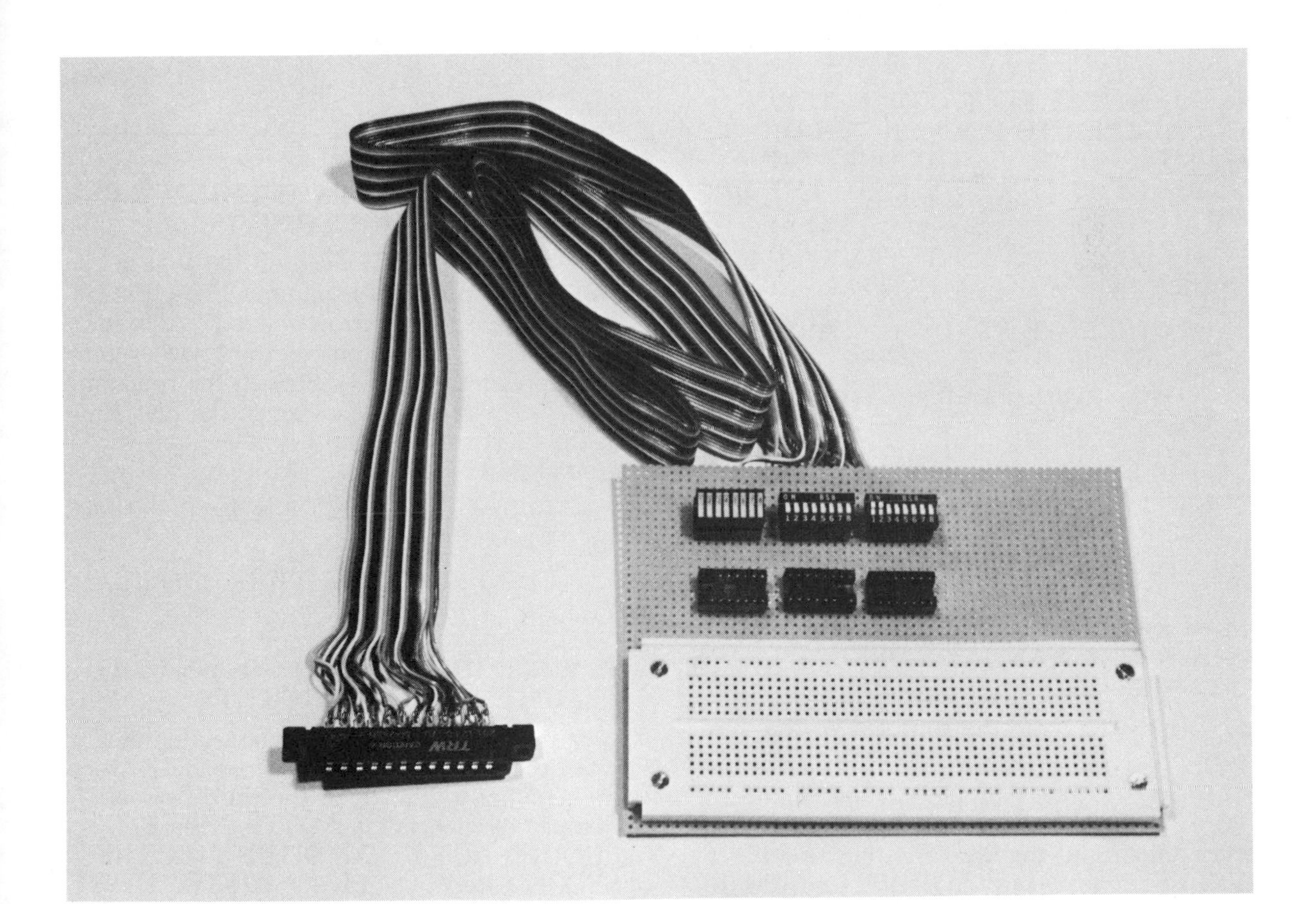

space as Pin 1. When we put the DIP switches into the sockets (step 8), the switches will be labeled (reading from left to right):

1-2-3-4-5-6-7-8-9-10-11-12-A-B-C-D-E-F-H-J-K-L-M-N

But, when you turn the board over (oriented with the sockets at the top), Pin 1 is in the far right position.

B) If this is not clear to you, read through the last two steps again before wiring the ribbon cable to the top row of leads on the DIP switch sockets. If you feel confident that you understand, look at your cable to determine which edge you wired to Pin 1 of the edge connector. This is the wire you will now connect to the *top, right* lead on the DIP switch sockets. (Remember, you turned the board over.)

IT MAY SEEM AS IF WE ARE USING TWO DIFFERENT PATTERNS FOR CONNECTING THE CABLES. BUT IF YOU LOOK CAREFULLY, YOU'LL SEE THAT PIN 1 ON THE EDGE CONNECTOR IS CONNECTED TO PIN 1 ON THE DIP SWITCH SOCKETS. THE ONLY DIFFERENCE IS THAT WE READ THE DIP SWITCHES:

1-2-3-4-5-6-7-8-9-10-11-12-A-B-C-D-E-F-H-J-K-L-M-N

WHILE WE READ THE CARD-EDGE CONNECTOR:

1-A-2-B-3-C-4-D-5-E-6-F-7-H-8-J-9-K-10-L-11-M-12-N

C) Connect wires using your wire wrap tool (#7). Follow the simple instructions on the package.

D) Wire wrap the *third wire* to the second lead. (REMEMBER THAT YOU ARE READING RIGHT TO LEFT AND THAT YOU WANT TO CONNECT LEADS 1-12 FIRST.)

E) Wire wrap the *fifth wire* to the third lead.

F) Continue wrapping the *odd* numbered wires around consecutive leads. If everything turns out right, the 23rd wire in the cable will be connected to the twelfth lead (Pin 12).

G) Now wire wrap the *even* numbered wires around the next 12 leads (A-N). For example, connect the second wire to the 13th lead.

H) Secure wires to leads with a little solder. Then, trim the long leads (first row only) back to the wire wrap.

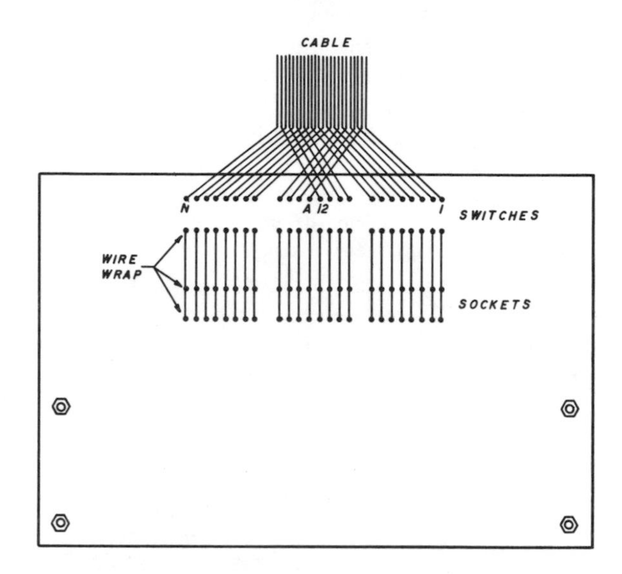

6. WIRE LEADS FROM THE JUMPER SOCKETS TOGETHER

A) Cut a 2-1/4" piece of 30-gauge wire (#8) and strip off all insulation. Using the wire wrapping tool, wrap about 1" of wire around the first lead on the third row (you can start at either end). Stretch the remaining 1-1/4" of wire straight down to the first lead in the fourth row. Wrap the remaining wire around that lead.

B) Repeat this procedure for the remaining 23 sets of posts.

C) DO NOT TRIM THE LONG LEADS YET.

7. WIRE DIP SWITCH SOCKETS TO JUMPER WIRE SOCKETS

A) Make a wire wrap connection between the leads in the second row (DIP switch sockets) and the leads in the third row (jumper wire sockets). NOTE: THERE SHOULD NOT BE ANYTHING ON THE SECOND ROW OF LEADS UNTIL THIS

POINT.

B) Cut a 2-1/4" piece of 30-gauge wire. Strip off all the insulation.

C) Wrap about an 1" of wire around the first lead in the second row (DIP switch sockets). Stretch the remaining wire straight down to the first lead in the third row. Wrap remaining wire around this lead.

D) Repeat this procedure for the remaining 23 sets of posts.

E) Solder each wire wrap to its lead so that the wire does not come loose.

F) Trim the long leads back to the wire wrap.

8. INSULATE EXPOSED WIRES

It is a good idea to insulate all the exposed wires and leads on the bottom of the Experimenter's Board. You might want to use a hot glue gun (see tool list page 40) to apply a coating of silicon "glue" to the wires and leads. (Don't worry, it is not sticky.) You could also buy some liquid silicon at a hardware store and paint the exposed wires and leads. Finally, you could cover the exposed wires and leads with tape. This is a somewhat clumsy technique, but it will work.

9. INSERT DIP SWITCHES INTO SOCKETS

A) Turn the board face up. Insert the three DIP switches in the top row of sockets. Position them with the "on" side at the top. NOTE: DIP SWITCHES VARY, BUT THERE IS USUALLY SOME MARKING TO INDICATE WHICH SIDE IS ON. IF THE SWITCH ITSELF IS NOT MARKED "ON", BUT THE LITTLE SWITCHES ARE NUMBERED, THE NUMBERS USUALLY INDICATE THE "ON" SIDE.

B) Carefully align the pins of the switches with the holes in the socket. Gently push each switch down onto its socket. Don't bend the pins.

10. CONNECT INTERFACE BOARD TO COMPUTER

This completes the assembly of your Experimenter's Interface Board. Remember, when you connect the card-edge socket to the Commodore 64's user port, keep the side marked "top" facing up.

BIBLIOGRAPHY

Here are some suggestions for additional reading:

ELECTRONICS

The Boy's Book of Radio and Electronics, Volumes 1 - 4. Alfred Morgan. New York: Charles Scribner's Sons, 1963-1969.

Electronics for Everybody. Ronald Benrey. New York: Harper and Row, 1970.

Electronics for Young People. (New Fifth Edition). Jeanne Bendick and R.J. Lefkowitz. New York: McGraw-Hill Book Company, 1973.

Getting Started in Electronics. Forrest M. Mims. Ft. Worth, Texas: Radio ShackTandy, 1983.

Transistors and Circuits: Electronics for Young Experimenters. W.E. Pearce and Aaron E. Klein. New York: Doubleday and Co., 1971.

Wires and Watts: Understanding and Using Electricity. Irwin Math. New York: Charles Scribner's Sons, 1981.

ROBOTICS

How to Build a Computer-Controlled Robot. Tod Loofbourrow. Rochelle Park, New Jersey: Hayden Book Company, Inc., 1978.

1984 Robotics Age Product Guide: A Sourcebook for Educators and Experimentalists. Order from Robotics Age Magazine, 174 Concord St., Peterborough, NH 03458; $9.95.

Rise of the Robots. George Sullivan. New York: Dodd, Mead and Company, 1971.

Robotics. Tony Potter and Ivor Guild. London: Usborne Publishing Ltd., 1983.

Robotics: Past, Present, and Future. David C. Knight. New York: William Morrow and Co., 1983.

Robots: Fact, Fiction, and Prediction. Jasia Reichardt. New York: The Viking Press, 1978.

Robots, Real to Reel. Barbara Krasnoff. New York: ARCO Publishing, 1982.

Super People: Who Will They Be? Jeanne Bendick. New York: McGraw-Hill Book Company, 1980.

The Robots Are Here. Alvin Silverstein and Virginia Silverstein. Englewood Cliffs, NJ: Prentice-Hall Inc., 1983.

Working Robot. Fred D'Ignazio. Hasbrouck Heights, NJ: Hayden Book Co., 1984.

MAGAZINES

Robotics Age Magazine. Order from Robotics Age, 174 Concord St., Peterborough, NH 03458.

Scientific American Magazine. Order from Scientific American, Inc., 415 Madison Ave., New York, NY 10017.

To help you in your search for parts and tools, especially as you get into more advanced projects, you might want to order free catalogs from the following electronics companies:

Active, P.O. Box 8000, Westborough, MA 01581

Advanced Computer Products, P.O. Box 17329, Irvine, CA 92713

ALI Electronics Corporation, 905 S. Vermont Avenue, P.O. Box 20406, Los Angeles, CA 90006

Chaney Electronics, P.O. Box 27038, Denver, CO 80227

Digi-Key Corporation, Highway 32 South, P.O. Box 677, Thief River Falls, MN 56701.

DoKay, 2100 De la Cruz Boulevard, Santa Clara, CA 95050

Heathkit, Heath Company, Benton Harbor, MI 49022

H & R Corp., 401 E. Erie Avenue, Philadelphia, PA 19134.

Mouser Electronics, 11433 Woodside Avenue, Santee, CA 92071.

Priority One Electronics, 9161 Deering Avenue, Chatsworth, CA 91311.

Radio Shack, over 6600 locations across the United States.

R & D Electronics, 1202H Pine Island Road, Cape Coral, FL 33909

R.F. Electronics, 1086-C N. State College Boulevard, Dept. R, Anaheim, CA 92806

Stock Drive Products, 55 S. Denton Avenue, New Hyde Park, NY 11040 (ask for Handbook of Small Standardized Components)

W.S. Jenks & Son, 2024 W. Virginia Avenue N.E., Washington, DC 20002

Here are two excellent sources for technical books on robotics:

H.D. Kohn Co., P.O. Box 16265, Alexandria, VA 22302

TAB Books, Inc., Blue Ridge Summit, PA 17214

SCHEMATIC DIAGRAMS

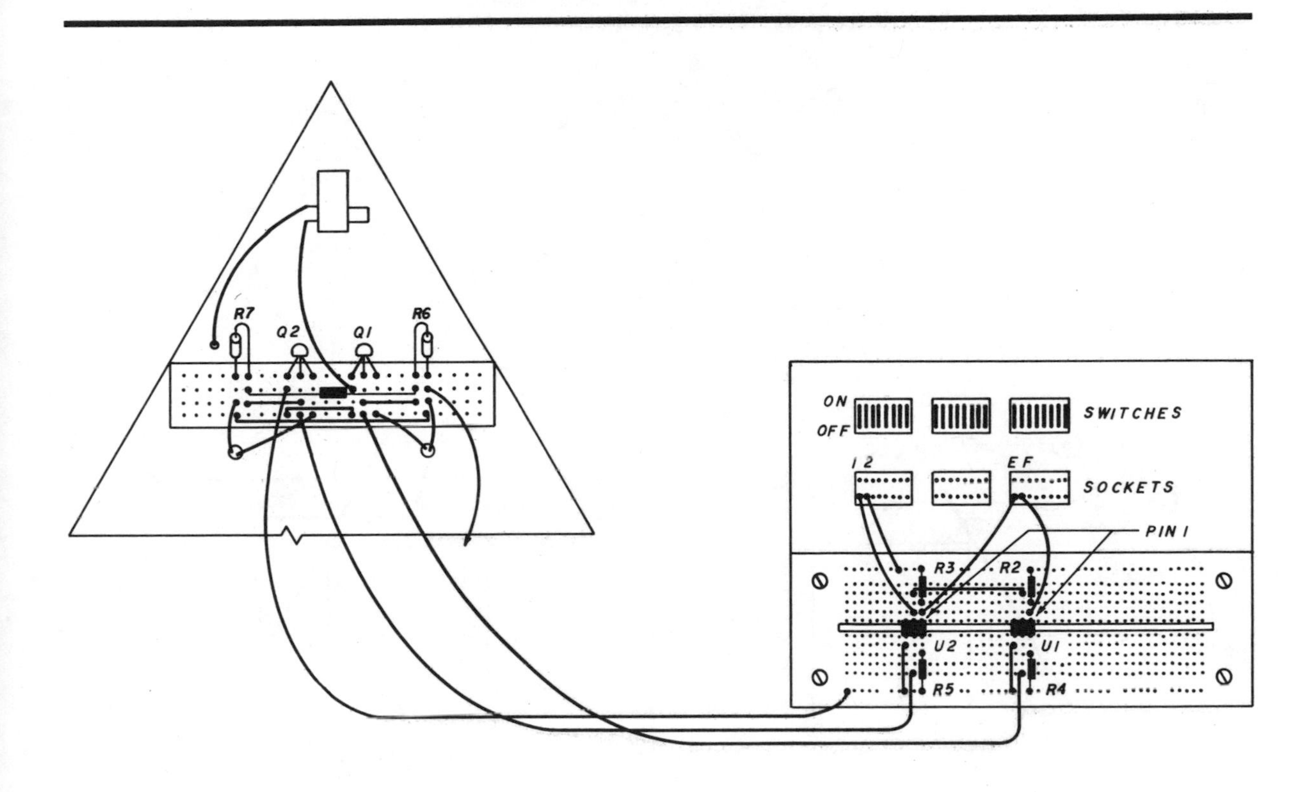

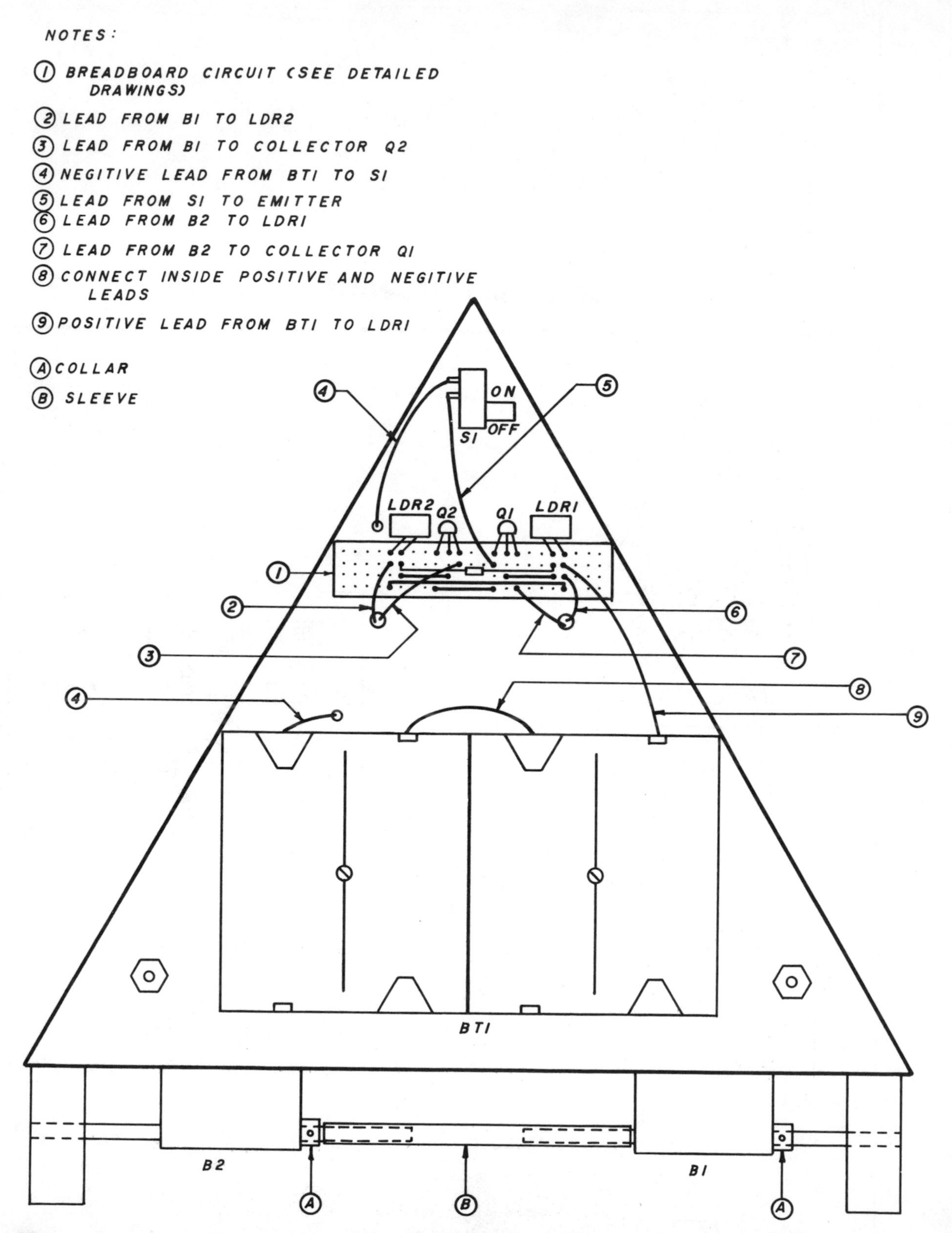
NOTES:
① BREADBOARD CIRCUIT (SEE DETAILED DRAWINGS)
② LEAD FROM B1 TO LDR2
③ LEAD FROM B1 TO COLLECTOR Q2
④ NEGITIVE LEAD FROM BT1 TO S1
⑤ LEAD FROM S1 TO EMITTER
⑥ LEAD FROM B2 TO LDR1
⑦ LEAD FROM B2 TO COLLECTOR Q1
⑧ CONNECT INSIDE POSITIVE AND NEGITIVE LEADS
⑨ POSITIVE LEAD FROM BT1 TO LDR1
Ⓐ COLLAR
Ⓑ SLEEVE
ON
OFF
S1
LDR2
Q2
Q1
LDR1
BT1
B2
B1